D1635231

BEHAVIOUR

Michael Dockery and *Michael Reiss*

CAMBRIDGE
UNIVERSITY PRESS

PUBLISHED BY THE PRESS SYNDICATE OF THE UNIVERSITY OF CAMBRIDGE
The Pitt Building, Trumpington Street, Cambridge CB2 1RP, United Kingdom

CAMBRIDGE UNIVERSITY PRESS
The Edinburgh Building, Cambridge CB2 2RU, United Kingdom
40 West 20th Street, New York, NY 10011-4211, USA
10 Stamford Road, Oakleigh, Melbourne 3166, Australia

© Cambridge University Press 1999

Typeset in Palatino 9.5pt [HM]

Printed in the United Kingdom at the University Press, Cambridge

A catalogue record for this book is available from the British Library

ISBN 0 521 59754 4 paperback

Dedication
To Martin Jones and Tim Clutton-Brock

ACKNOWLEDGEMENTS

The authors are particularly grateful to Dr Mike Hansell who read through the entire manuscript and suggested valuable improvements, to Diane Abbott, Aidan Gill and Sue Kearsey, our patient, helpful and efficient editors, and to Janice Ainsworth for help with the typing.

The authors and publisher would like to thank the following for permission to reproduce their material.

Photos
1.2 Roger Tidman/NHPA; 2.2 Ben Osborne/Oxford Scientific Films; 2.5 Royal Society for the Protection of Birds; 2.6l M. Gilroy/Aquila; 2.6r H. Lacey/FLPA; 3.1 Tony Tilford/Oxford Scientific Films; 3.2 Roger Brown/Oxford Scientific Films; 4.4 Professor N.J. Mackintosh, University of Cambridge; 5.4 David Macdonald/Oxford Scientific Films; 5.5tl David G. Fox/Oxford Scientific Films; 5.5tr Jane Burton/Bruce Coleman Ltd; 5.5bl, 5.5br K.G. Preston-Mafham/Premaphotos Wildlife; 5.6 John Watkins/FLPA; 6.2 John Cancalosi/BBC Natural History Unit; 6.7 Antony Medley/SIN; 7.2 Stephen Dalton/Oxford Scientific Films; 7.4 Raymond Mendez/Animals Animals; 7.6 Anup Shah/BBC Natural History Unit.

Contents

1	**The science of behaviour**	1
1.1	Introduction	1
1.2	Ethology, anthropology, psychology, sociology and sociobiology	2
1.3	Tinbergen's four 'Why's'	3
1.4	The functions of behaviour	4
1.5	The unit of natural selection	8
1.6	Behavioural genetics	9
1.7	The evolution of behaviour	10
2	**The development of behaviour**	13
2.1	Nature/nurture	13
2.2	Instinctive behaviour	17
2.3	Attachment and imprinting	19
2.4	Play	24
3	**Responding to the environment**	28
3.1	Detecting stimuli	28
3.2	Processing information	29
3.3	Principles of communication	31
3.4	Territorial advertisement	33
3.5	Communication to resolve conflicts	38
3.6	The bee language controversy	40
3.7	Language	43
4	**Learning**	44
4.1	Habituation	45
4.2	Classical conditioning	46
4.3	Operant conditioning	48
4.4	Latent learning	51
4.5	Insight learning	52
4.6	Observational learning	53
4.7	Memory	54
4.8	Intelligence	56
5	**Obtaining food and avoiding being eaten**	59
5.1	Optimal foraging	60
5.2	Factors influencing foraging decisions	64
5.3	Strategies used by animals searching and hunting for food	65
5.4	Selecting what to eat	67
5.5	Avoiding attack by predators	68
5.6	Avoiding capture by predators	71

Contents

6	**Courtship and mating behaviour**	74
6.1	Life in a gull colony – an example of courtship	74
6.2	Must there always be a father?	75
6.3	How do males compete?	77
6.4	Female competition and selection of mates	83
6.5	Female–male conflict	86
6.6	Mating systems	88
6.7	Human courtship and reproduction	90
7	**Social behaviour**	96
7.1	Parental behaviour	96
7.2	Costs and benefits of being social	97
7.3	Evolution of altruism	100
7.4	The social life of insects	104
7.5	The social life of African wild dogs	109
7.6	Chimpanzee society	110
7.7	Human society	112
	Further reading and videos	113
	Index	115

ACKNOWLEDGEMENTS (continued)

The authors and publisher would like to thank the following for permission to reproduce their material.

Artwork

1.1 based on figure 10.5 R.A. Hinde (1974) *Biological bases of human behaviour*, by permission of the McGraw-Hill Companies, New York; 1.3 based on page 199 N. Tinbergen (1974) *Curious Naturalists* (revised edn.) Penguin, by permission of Reed International Ltd; 1.4, Dr Jenny Chapman; 2.1 based on figure in *Animal Behaviour* **11**, A.W. Ewing 369–378 (1963) by permission of the publisher Academic Press; 2.7 based on figure on page 114 *Animal Behaviour* **52**, K.V. Thompson Play-partner preferences and the function of social play in infant sable antelope, *Hippotragus niger*. 114–15, (1996) by permission of the publisher Academic Press; 3.3 based on figures on pages 63, 64 and 65, A. Gosler (1993) *The Great Tit*, by permission of A.G. Gosler and Norman Arlott; 3.4 based on figure 6.4, J.R Krebs & N.B. Davies (eds.) (1984) *Behavioural Ecology: an Evolutionary Approach*, 2nd edn. Blackwell Science Ltd, Oxford; 3.6, 3.7 M. Reiss & H. Sants (1987) *Behaviour and social organisation*, Cambridge University Press. 4.1 based on figure on page 15, John Sparks (1969), *Bird Behaviour*, © copyright 1969 by Hamlyn Publishing Group Ltd; 4.2 based on Hilgard *et al. Introduction to psychology* (7th ed.) © 1979 Harcourt Brace Jovanovich; 4.6 based on figure on page 62, J.L. Gould & C.G. Gould (1994) *The Animal Mind*, Scientific American Library, Macmillan Distribution Ltd; 5.2, 5.3 based on figures 1 and 2, *Animal Behaviour* **51**, M.H Persons & G.W Uetz, The influence of sensory information on patch residence time in wolf spiders (Aranae: Lycisidae), 1285–93, (1996) by permission of the publisher Academic Press; 5.7 based on figure 4.2a J.R. Krebs (1982) *An Introduction to Behavioural Ecology* and R.E. Kenward (1978) *J. Anim. Ecol.* **47**, 449–460, by permission of Blackwell Science Ltd; 6.3 based on figure 1, *Animal Behaviour* **40**, J. Höglund M. Erickson & L.E. Lindell, Females of the lek-breeding great snipe, *Gallinago media*, prefer males with white tails, 23–32, (1990) by permission of the publisher Academic Press; 6.4 based on figure 1, *Animal Behaviour* **40**, J. Höglund & G.M. Robertson, Female preferences, male decision rules and the evolution of leks in the great snipe, *Gallinago media*, 15–22, (1990) by permission of the publisher Academic Press; 6.5 based on figure 2, *Animal Behaviour* **37**, J.H. Poole, Mate guarding, reproductive success and female choice in African elephants, 842–49, (1989) by permission of the publisher Academic Press; 6.6 reprinted from *Trends in Ecology and Evolution*, **1**, M.E.N. Majerus, The genetics and evolution of female choice, pages 1–7, © 1986, with permission from Elsevier Science; 7.1 based on figure from B.W. Sweeney 7 R.L Vannote (1982) Population synchrony in mayflies: a predator satiation hypothsesis. *Evolution*, **36**, 810–821.

Table

2.1 reprinted from *Animal Behaviour* **49**, T.M. Caro, Short-term costs and correlates of play in cheetahs, 333–345 (1995), by permission of the publisher Academic Press.

Every effort has been made to reach copyright holders. The publisher would be glad to hear from anyone whose rights they have unknowingly infringed.

ONE

The science of behaviour

1.1 Introduction

This book is concerned with the science of behaviour. Much of the material is to do with non-human animals but we have also included some consideration of human behaviour. Indeed, throughout the book 'animals' includes humans. We have attempted to show how an understanding of the behaviour of non-humans can help us to understand ourselves better. At the same time, although there are similarities between humans and non-humans, there are very significant differences too. These differences mean that we should be careful not to extrapolate too readily from the behaviour of non-humans to our own behaviour. Equally, we should be careful not to be **anthropomorphic**. To be anthropomorphic is to generalise from humans to non-humans. For example, you might be tempted to think that the rhesus macaque in figure 1.1 is grinning because it is happy. Actually, careful observation shows that quite the reverse is the case. This 'grin' is better named as a 'bared-teeth display'. (Such a label describes what the behaviour looks like, not what its function is guessed to be.) In fact, the behaviour in figure 1.1 is displayed when the macaque would like to flee but is thwarted from so doing – for example, because it is cornered.

Figure 1.1   The bared-teeth display of a rhesus macaque. It may look as though the animal is grinning happily. Actually, it is fearful.

While anthropomorphism is not good science, neither is its extreme opposite. Humans, while distinct from other animals, are not utterly different from them. We can learn about ourselves by studying the behaviour of other species.

In this chapter we will look at some of the questions which the science of behaviour attempts to answer and will also outline how those who study behaviour attempt to answer such questions. We begin by looking at the various scientific disciplines that are concerned with the study of behaviour.

1.2 Ethology, anthropology, psychology, sociology and sociobiology

There are a number of different scientific disciplines which look at behaviour. Actually, the boundaries between these various disciplines have shifted over time and nowadays are not as rigid as they once were. Nevertheless, each of these disciplines has its particular focus and methodology.

Ethology is the study of the behaviour of animals – mainly non-humans. The golden days of ethology were from the 1920s to the 1970s and we shall refer in this book to some of the great ethologists such as Konrad Lorenz, Niko Tinbergen, Karl von Frisch and Jane Goodall. The fundamental working assumption of ethologists is that in order to understand behaviour, animals should be studied carefully in their natural environment. Experiments are often performed and are undertaken in the field not in a laboratory. Little equipment is needed beyond a notepad, a pencil and a watch, and the key to success is patience and acute observations. Often observations are carried out over a period of many years, allowing a detailed understanding of the behaviour of the species to be built up gradually.

Anthropology is the study of living people. It has several branches. Cultural anthropologists study the cultures of living peoples. They examine what people make (cultural artefacts), the ways they communicate and how they organise their lives. Traditionally, a cultural anthropologist would spend one or more years away from home living with a foreign people. Many cultural anthropologists were especially interested in peoples whose lives had changed little in recent centuries. So they would study tribespeople, peasants and others relatively unaffected by industrialisation and westernisation. Nowadays, cultural anthropologists, while continuing to study such people, also study cultures nearer to home. There are urban anthropologists who study how city dwelling affects people, and feminist anthropologists who focus their research on the roles occupied by women in culture and society. There are other branches of anthropology. For example, physical anthropologists are interested in how humans evolved. What were the evolutionary pressures that led to our having such large brains, to standing upright and to having so little body hair? How have our bodies adapted to the many environments in which we live?

Psychology looks both at humans and at non-humans. Psychologists are interested in understanding how animals think and why they behave as they do. Often the emphasis has been on the mechanisms underlying behaviours. How does learning take place, for example? And what precisely is involved in 'seeing'? The answer to this question requires both an understanding of the physiology of the eye and an understanding of how the brain processes the information gathered by an animal's eyes. Traditionally, psychologists studied their subjects under laboratory conditions so that they could precisely control the environment. This allowed psychologists to investigate one question at a time. For example, when an animal is shown a number of objects for a brief period of time, what determines what the animal subsequently remembers? Nowadays psychologists, while continuing to work in the laboratory, also work in less controlled environments.

Sociology is the study of how humans live in society and behave in groups. These groups may vary from the temporary groups formed such as when people are at a sporting event to the longer-term groups that occur at home and in work. A sociologist might study how a person's educational achievements and earning potential are affected by their social class – that is, by the type of job that they have. Or a sociologist might be interested in how different societies are structured by economic, political and other factors. Sociologists are also interested in how people differ in their access to power and in how they make use of their work and leisure time.

Sociobiology burst onto the scene in 1975 when E. O. Wilson produced a book of this title. Sociobiology concentrates on the social behaviour of animals, examining the reasons why animals have evolved to live and behave in groups as they do. Social behaviour often involves helping behaviours, and sociobiologists are particularly interested in explaining the advantages and disadvantages of helping behaviours. They are also interested in understanding the genetic basis of behaviours. In the late 1970s and early 1980s many sociobiologists, including E. O. Wilson himself, enthusiastically extrapolated from the behaviour of non-humans to the behaviour of humans. Sociobiologists argued that many of the differences we see typically between men and women – for example that men are more likely to be aggressive and women more likely to bring up children – were primarily due not to culture but to our genes. This caused tremendous controversy and sociobiologists were often accused of being sexist or racist.

1.3 Tinbergen's four 'Why's'

The Greek philosopher, Aristotle, pointed out that four *causes* can be identified for most objects or events. What causes a house, for example? One answer is the matter from which it is constructed; a second is the builder; a third is the plan of the house; and a fourth is the house's purpose. Similarly, the Dutch ethologist Niko Tinbergen realised that when people ask *why* a certain behaviour occurs, they may mean one or more of four things. They may mean:

- what are the mechanisms that enable that behaviour to occur?
- how did the behaviour develop during the life of the individual showing it?
- what is the function of the behaviour?
- how did it evolve over the generations?

Consider, for example, a blackbird singing. We can ask what are the mechanisms that enable singing to take place. (The answer will have something to do with vocal cords and breathing and muscles and nervous control.) Or we can ask how the behaviour develops as a blackbird grows up. (To answer this question we might try tape-recording a blackbird's song to see whether it changes during an individual's life. We might also see whether blackbirds need to hear other blackbird songs before they can sing themselves.) Then we can ask what the function of the blackbird song is. (Is it to attract a mating partner, for example, or is it to proclaim ownership of a territory to rivals?) Finally, we can ask how blackbird song evolved.

The first of Tinbergen's four questions is really to do with physiology and we shall say only a little about it in this book. The last of Tinbergen's questions is the most difficult of the four to answer and we shall say only a little about the evolution of behaviour here. Most of what we cover is concerned with the functions and development of behaviour.

1.4   The functions of behaviour

We shall illustrate the ways in which scientists have investigated the functions of behaviour by considering two of the classic stories of animal behaviour – eggshell removal in black-headed gulls and the reproductive behaviour of sticklebacks.

Eggshell removal in black-headed gulls

During the 1950s and 1960s, Tinbergen led a large research programme into the behaviour of black-headed gulls, herring gulls and kittiwakes. The work on black-headed gulls was done in Cumbria in the north-west of England. Black-headed gulls (*Larus ridibundus*) are social breeders. They crowd their nests together even when apparently suitable nest sites are available elsewhere. The birds are mainly **monogamous**, so that a single female and a single male form a pair. Within the colony there is a system of **territories**, each territory being an area defended by a single pair of gulls (figure 1.2).

Early in spring, the birds gradually return to the breeding colony from their winter quarters. Some birds arrive paired, others only pair up after they arrive at the colony. Once the birds arrive, much calling and posturing

Figure 1.2 Black-headed gulls (*Larus ridibundus*) defending territories at their breeding colony.

occur as the birds establish territorial boundaries. When a bird intrudes into another pair's territory, the territory owners, particularly the male bird, usually respond by posturing and calling but occasionally by attacking the intruding bird. Typically the intruding bird retreats. Actual fights are rare.

After all the courtship and territory establishment is over, eggs are laid, incubated and eventually hatched. Then, within a few hours of a chick hatching, a small but intriguing piece of behaviour occurs. One of the parents takes the empty shell in its bill, walks or flies away with it, and drops it well away from the nest. It took Tinbergen a long time to become interested in this because it seemed such a minor piece of behaviour. However, as Tinbergen spent longer observing the birds he noticed that predators such as neighbouring black-headed gulls, marauding herring gulls or passing carrion crows were often on the alert for just such occasions when a gull leaves its brood unprotected for a few seconds. The gulls or crows break open eggs or grab and eat a chick. Evidently, eggshell removal has a cost. But what is its corresponding benefit?

Tinbergen came up with a number of rival hypotheses for the function of eggshell removal:

- eggshell removal prevents the sharp edges of an eggshell from injuring the newborn chick (such damage had occasionally been reported in duck hatcheries);
- eggshell removal prevents an empty shell from slipping over an, as yet, unhatched egg, imprisoning the chick inside;
- in the absence of eggshell removal, an empty shell might compete for one of the gull's three brood spots – the defeathered patches on its belly used for incubation;
- disease-causing bacteria grow in empty eggshells;
- empty eggshells have white interiors and attract the attention of predators.

Tinbergen thought the last of these five explanations especially worth testing for two reasons. First, many predators, including carrion crows and foxes, do take black-headed gull eggs and chicks. Secondly, the kittiwake, a bird closely related to the black-headed gull, rarely removes its eggshells. Kittiwakes nest on very steep cliffs where predation is much less of a problem than with black-headed gulls. Indeed, kittiwake chicks don't even have the camouflaged plumage typical of gull species. Instead of the buff ground colour and speckled dark dots typical in gulls, kittiwake chicks are a beautiful silver-white.

To test whether eggshell removal reduces predation in black-headed gulls, Tinbergen laid out a number of well-scattered, single eggs in a valley next to an area where the gulls nested. Half these eggs were left 'isolated'; the others each had an empty eggshell placed just 5 cm from them. Tinbergen and his co-workers then retired to a hide on a nearby dune-top. Carrion crows, herring gulls and an occasional black-headed gull swooped down and attacked the intact eggs. When about half the eggs had been taken, counts were made to determine which eggs had survived. Many more of the 'isolated' eggs survived. When the effect of distance between the eggs and the eggshells was examined more systematically, a clear pattern emerged (figure 1.3). The greater the distance between an intact egg and a broken eggshell, the lower the chance that a predator would take the intact egg.

Observations of the birds that fed on these eggs showed why eggshell removal is advantageous. On spotting an empty eggshell from afar, a crow or gull alights near to, or walks up to, the empty shell and then starts searching around it. The closer the intact egg, the more likely it is to be found. So here is evidence for at least one function of eggshell removal – it reduces the risk of predation.

Tinbergen's work on eggshell removal is seen as a landmark investigation in *ethology*. He presented clear alternative hypotheses, explicitly considered the costs and benefits of a behaviour, and experimentally tested his ideas.

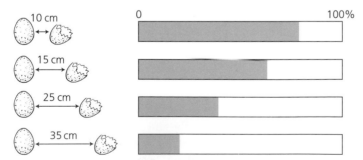

Figure 1.3   The results of Tinbergen's experiment to determine the effect of broken eggshells on the predation of intact eggs. The closer an eggshell is to an intact egg, the greater the risk of predation. The lengths of the shaded parts of the bars show the percentage of eggs taken, within a standard time, for each of the egg-to-eggshell distances indicated.

Reproductive behaviour in sticklebacks

Sticklebacks are small freshwater fish. In a series of experiments dating from the 1930s, Tinbergen investigated the reproductive behaviour of the three-spined stickleback (*Gasterosteus aculeatus*). Tinbergen's account can be summarised as follows (figure 1.4).

In spring, male sticklebacks set up territories from which they chase away intruders of either sex. At the same time, each male constructs a nest and develops a red belly. The red on a male's belly acts as a **sign stimulus** – that is, it provokes a **stereotyped response**, in this case aggression from a territorial male. Tinbergen found that a realistically shaped but non-red model male stickleback provoked little interest from a territorial male, whereas extremely crude models painted red on their lower surfaces provoked strong aggression!

Once a male's nest is complete, the male becomes interested in females swollen with eggs. When a female appears, he moves towards her in a curious zig-zag fashion (the zig-zag dance). When she sees him, the female responds by swimming towards the male with her head and tail turned upwards, thereby displaying her swollen abdomen. The swollen abdomen acts as a sign stimulus for the male. Realistic model females lacking a swollen belly are not courted, while crude model females provided with a swollen lower surface are.

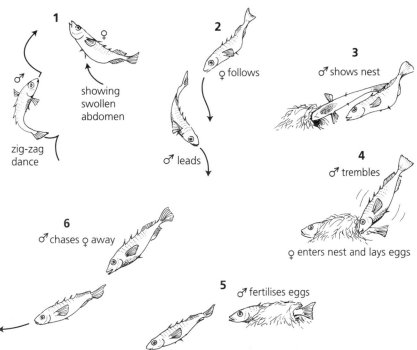

Figure 1.4   Courtship and reproduction in the three-spined stickleback.

Having displayed her swollen abdomen to the male, he then leads her to the nest entrance and shows it by poking it with his snout. The female enters the nest and the male gives her several prods with a trembling motion to stimulate her to lay her eggs. When she has discharged all her eggs, she leaves the nest and the male enters it and ejaculates sperm over the eggs. He then chases the female away.

A male may mate with as many as five different females. Then, having fertilised several clutches, he loses his readiness to court females. Instead he begins regular ventilation of the eggs by fanning them with his pectoral fins. The time spent in fanning increases daily until the eggs hatch, when it stops. The fanning is to aerate the eggs. Evidence for this is provided by the observation that artificial lowering of the oxygen content of the water induces increased fanning.

The above account was the standard one found in animal textbooks for almost half a century. Then, in the 1980s, the story became modified in a number of significant ways which provide valuable lessons about scientific research. Perhaps the most remarkable finding is that careful experiments have shown that adding red to a model stickleback male makes it *less* likely to be attacked by a territorial male! This, of course, is the exact opposite of what Tinbergen wrote. We can't be certain why Tinbergen got it wrong. Possibly he was so sure what he would find that he failed to record adequately those occasions when his experiment 'didn't work' – that is when a model without red was attacked more than a model with red. In any event, the whole saga emphasises the importance of making and keeping proper records and of testing hypotheses statistically – something that Tinbergen did not do.

Another way in which Tinbergen's account has been modified came about through researchers studying females in the same ways as they studied males. In the wild, females go around in groups. Experimental studies of such all-female groups have shown that there are actually more aggressive encounters per hour in all-female groups than in all-male groups. What is the function of this female aggression? It turns out that females compete amongst themselves for access to males with the best territories. In other words, females cannot be viewed, as Tinbergen had unthinkingly assumed, as the passive recipients of a male's attentions. They too have reproductive interests. Indeed, as we shall see (section 6.4), by and large females in the animal kingdom are more choosy than males about their mating partners.

1.5   The unit of natural selection

You may be used to thinking of natural selection as acting on individuals. For example, one form of a moth may survive and reproduce more successfully than another form. We can talk about some individuals being **fitter** than others, that is, leaving more descendants.

However, there are occasions, as even Charles Darwin realised, when individuals clearly behave in ways that do not maximise their individual fitness. Consider the tens of thousands of worker honey bees in a single colony. None of these workers ever reproduces. Instead they devote their entire lives to helping rear the offspring of the queen. What is the evolutionary explanation for this behaviour? The simple answer is that they are helping to produce not their offspring but their sibs, that is their sisters and brothers. The worker bees themselves are the daughters of the queen. In helping the queen to rear offspring, they are therefore helping their mother to have sons and daughters. These sons and daughters are their sibs. Darwin termed this 'family selection'. Nowadays it is called 'kin selection' and we shall examine it in more detail in section 7.3. For the moment, we shall introduce one more technical term – namely **inclusive fitness**. The idea is that an individual can, in a sense, reproduce via its relatives, as we saw in our example of a worker honey bee. The term 'inclusive fitness' therefore takes account of the helping behaviour an individual gives its relatives. In other words, it is not an organism's individual reproductive success alone that matters, but also the reproductive success of any of its relatives that have been affected by its helping behaviour.

The notion of inclusive fitness is quite a complicated one. Many evolutionary biologists, notably Richard Dawkins, argue that the best way of understanding evolution is not to think about individuals having offspring or even helping relatives to, but rather to think of genes replicating and leaving copies of themselves. You may well have heard of the term 'the selfish gene'. It comes from a book of that title written by Richard Dawkins which is well worth reading.

1.6   Behavioural genetics

If behaviour is to evolve by natural selection, then behavioural differences between individuals need to have a genetic basis. Evidence that behaviours may have a genetic component to their inheritance comes from the many different breeds of domestic dog. We now think that dogs were first domesticated some 140 000 years ago. Most breeds, though, are less than a thousand years old. Even a thousand years, however, has been sufficient time for several hundred generations and this has allowed considerable differences in behaviour to evolve under artificial selection. Think, for example, of the way different breeds of dogs behave towards strangers or react to an object being dropped into water. Even puppies of the different breeds behave very differently, suggesting that the differences are genetic rather than due to training or other aspects of the environment.

One of the earliest demonstrations of the genetic basis of a behaviour involved the nest-clearing behaviour of honey bee workers. Bees in some hives have the ability to perform two behaviours:

- first, uncapping cells that contain dead pupae;
- secondly, removing the dead pupae from the cells and the hive.

Workers from other hives lack these abilities and ignore cells with dead pupae. Rothenbuhler crossed the two strains and then performed some additional crosses with the hybrid generation. These experiments revealed that the expression of each of the two behaviours was under the control of separate genes called u (for uncap) and r (for remove). Workers that were homozygous recessive for each gene (i.e. uurr) would first uncap cells with dead pupae and then remove them. Workers with one or two copies of the dominant allele of each gene (i.e. UURR, UURr, UuRR or UuRr) would only rarely uncap or remove. Through his genetic crosses Rothenbuhler produced some bees with the genotype Uurr. These were unable to uncap cells but would remove dead pupae provided Rothenbuhler did the uncapping for them!

The above nest-clearing story is an elegant one. However, it is beguiling in its simplicity for at least two reasons. First of all, most behaviours, certainly human ones, are not under the influence of just one or two genes. Any genetic component is much more likely to be polygenic (that is affected by the actions of many genes). Secondly, as we shall see in section 2.1, it is more fruitful to think of behaviours as being the result of an interaction between genetic and environmental effects.

1.7   The evolution of behaviour

Already in our consideration of eggshell removal in birds, we have found ourselves thinking about the differences between black-headed gulls and kittiwakes and so hypothesising on the evolutionary history of the behaviour. A full understanding of a behaviour really requires answers to all four of Tinbergen's questions: What is the function of the behaviour? How does it develop? What mechanisms underlie it? How did it evolve? (section 1.3).

Behaviours do not fossilise in the way that bones do. True, the discovery of fossilised dinosaur nests tells us something about the incubation behaviour of the adults. Similarly, animals sometimes leave behind **trace fossils** as a result of their movements or the homes they have constructed. Nevertheless, the study of the evolution of behaviour relies heavily on the **comparative method**. The comparative method works by comparing the behaviour of related living species and then trying to piece together a possible evolutionary pathway for the behaviour. It is not the most experimental of sciences, though it can make testable predictions.

An elegant example of the use of the comparative method is provided by the study of how a distinctive courtship signal evolved in a small empid fly, *Hilara sartor*. The adult male of this species constructs a hollow silk

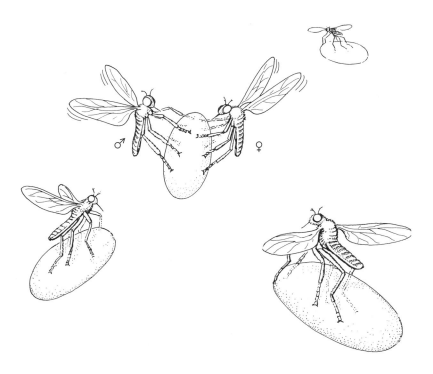

Figure 1.5 Courtship in the fly *Hilara sartor*. A swarm of four males is shown. A female (♀) is in the act of accepting an empty silk balloon from one of the males (♂)as a precondition for mating.

balloon that is almost as large as he is. He then flies to a swarm composed of other balloon-carrying males, and there circles about until a female arrives (figure 1.5). The female selects a male from the swarm, accepts his balloon and the two leave the swarm and mate.

So how did this behaviour evolve? Without comparative data the question might remain unanswered. However, there are thousands of species of empid flies related to *H. sartor*. Kessel has summarised the available information and divides empids into eight groups on the basis of their reproductive behaviour. The eight groups are thought to provide an evolutionary sequence from the ancestral condition (in group A) to *H. sartor* and a few other empids (in group H).

- Flies in group A are carnivorous, hunting for other small flies. Males in group A search for a female and court her in isolation from other males.
- Males in group B capture a prey item just before locating a female. The female takes the food and consumes it during mating.
- Males in group C capture a prey item and then form swarms with other courting males. A female attracted to the swarm selects a male, receives the prey and mates.

- Males in group D behave the same as males in group C except that they apply some strands of silk to the prey before joining the swarm.
- Males in group E behave as males in group D except that they wrap the prey item entirely in a heavy silk 'bandage' before joining the swarm.
- Males in group F behave as the males in group E except that they remove the juices from their offering before wrapping it. As a result, each female receives a non-nutritious husk.
- Males in group G feed only on nectar. However, a courting male finds a dried insect fragment and uses it as a foundation for the construction of a large silk balloon before joining the swarm.
- Males in group H don't even include the insect fragment in their balloon.

Assuming that the above evolutionary sequence is correct, we can speculate about the selection pressures that led to these behaviours. Kessel and others have suggested that males in group B offer a 'gift' because of the predatory nature of the female and consequent risk to the male of cannibalism by the female. The present serves as a distraction, enabling the male to get on with reproduction more safely while the female eats the prey. An alternative, and not mutually exclusive, hypothesis is that the gift provides the female with a high protein meal that helps her to lay more eggs. Thus it is to her advantage to choose a gift-bearing male and, in turn, to the male's advantage to offer such a gift. Perhaps the continued existence of group A courtship in some empid species stems from the failure of any mutant male in these species to offer a female a prey item while courting. Or perhaps there is something about their ecology that reduces the risk of cannibalism or reduces the advantage to a male of offering the female a prey item for a meal.

The same kind of speculation can be applied to every other change. Gathering in swarms may make males more conspicuous to females and so may be to the advantage of successful males. Once some males swarm, it may be to a female's advantage to ignore solitary males, instead visiting only swarms as these allow her to compare males before mating with one of them (group C). The silk may initially serve to retain prey (group D) and then to make it look larger than it really is (group E). By this stage, females accept males on the characteristics of their silk balloon which allows a male to 'cheat' by consuming the prey himself before wrapping it (group F). Upon adoption of a purely nectivorous feeding niche (group G), prey capture serves no nutritional function, but continues, perhaps as a vestigial behaviour. Finally, prey capture itself is lost (group H).

Unravelling the evolution of courtship in this fly relied simply on observations of behaviour. Nowadays, direct DNA sequencing can often be combined with information on anatomy and behaviour to predict evolutionary pathways.

TWO

The development of behaviour

2.1 Nature/nurture

As a young organism grows its body changes – for example, its mass increases and its limbs lengthen. The behaviour of an organism changes too. How is this controlled? Some behaviours, especially those associated with the very early stages of an individual's life, develop without any apparent influence of the environment or the experience of the organism – the behaviour develops due to the **nature** of the organism. This means that the behaviour is set in train at the appropriate time by some internal mechanism and that the conditions in which the organism is growing are not influential. This implies that the behaviour arises from inherited characteristics – that is, the behaviour is innate and is already built into the organism at fertilisation. On the other hand, some behaviour is environmentally determined – that is, the organism behaves in an appropriate manner as a result of its experience in the environment in which it lives. Such behaviours are said to be due to **nurture**. An organism's experience might be gained through interacting with its parents and sibs, others in its group, with predatory animals, with the food available in the environment, and so on.

In this chapter we are looking at how nature and nurture contribute to the development of behaviour and how they interact.

How genes affect behaviour

A number of studies have shown a genetic influence on behaviour. You have already met one example in section 1.6, relating to hygienic and unhygienic strains of honey bees. Another study on genetic influence was carried out on rats. In this experiment the researcher put a number of rats through a maze and was interested in determining if they differed in their ability to learn their way through it. Each animal had to explore the alleys in order to reach the goal, or end point, which was a food box at the end of the maze. Not surprisingly, it was found that the rats varied in how quickly they reached the goal and this allowed the researcher to identify those that went through quickly, which were called the 'maze bright' rats, and those that went through slowly, called the 'maze dull' rats. Individuals in these two groups were separated, allowed to breed with other rats in their group, and their offspring were then tested. By the time six generations had passed it was found that the 'dull' rats made roughly twice as many mistakes as the 'bright' rats. The difference in ability between the two groups clearly had a genetic basis which could be inherited. So this example also demonstrates how nature and nurture interact, since it was the rats' *learning ability* that was inherited. In follow-up experiments with different mazes it was found that the performance of the 'bright' and 'dull' rats did not differ, which shows that the efficiency of maze learning varies from maze to maze. It is therefore difficult to select for a general skill, such as 'brightness' in rats.

In an experiment with fruit flies (*Drosophila melanogaster*), a researcher tried to select for flies that differed in rates of activity. He arranged for flies to be placed in apparatus consisting of five funnels linked together (figure 2.1). The researcher found that some flies moved through the apparatus speedily, some slowly and some at an intermediate rate. Individuals from the first two groups were kept and allowed to breed – slow flies with slow flies, fast flies with fast flies. He found that the difference in speed between the two groups increased over the generations. Interestingly, however, he found that there were no apparent differences, even after many generations, between the flies from each group when they were placed in an open area. What the experiments seemed to have selected for was not speed but tolerance to crowding. In the crowded experimental set-up of figure 2.1, fast flies were rapidly trying to escape from the presence of the other flies. These experiments show how artificial selection in the laboratory can act on a specific behavioural trait. They also show how experimental results often turn out to be different from those anticipated by a researcher.

Of course, if there are genetic influences on behaviour they are seldom the result of one particular gene, rather many genes are involved. One way to look at how groups of genes can influence behaviour is to look at how hybrid offspring from two closely related species behave and contrast this with the behaviour of individuals of each parent species. Lovebirds were used in one well-known study, which looked at how males carry material to their nest site. Two species were involved, and the aim of the study was to investigate how this behaviour developed in hybrid male lovebirds. Males

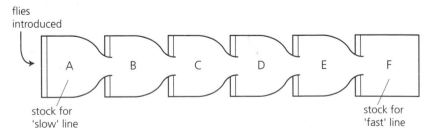

Figure 2.1 Apparatus used in the experiment with fruit flies. Fifty flies were placed in the apparatus: the last ten to leave the first funnel made up the 'slow' population and the first ten to reach the final funnel made up the 'fast' population.

of one species (*Agapornis personata*) carry the strips of material in their beaks; males of the other species (*Agapornis roseicollis*) tuck the strips under a wing. The hybrid males showed elements of both species-typical behaviours, sometimes in long disorganised sequences which resulted in failure. This shows that genetic influences from both species were acting but were in conflict, not leading to a smooth co-ordinated sequence of behaviour to produce a viable nest. However, after several months hybrid males did become more efficient at tucking material under their feathers and flying with it to the nest. This illustrates that experience can also influence the development of the behaviour.

Experiments in artificial selection have been conducted for many years as humans have tried to increase the likelihood of certain animals exhibiting particular behaviours. Breeding racehorses is just one example. This example neatly illustrates how 'hit and miss' breeding for a particular behaviour often is – mating two Classic winners does not necessarily produce offspring which go on to win subsequent Classic races. This is eloquent witness to the considerable genetic variation in organisms, and that running fast is a highly complex behaviour that is under the control and interaction of many genes. It means that, since many genes are involved, it is likely to be very difficult to achieve the best combination of genetic charac-teristics to produce a 'winner'. Running fast is, however, also under the influence of environmental factors, such as the quality of the food and the training a horse receives.

How environment affects behaviour

Even inside an egg or a womb a growing organism can respond to environ-mental stimuli. Once an animal is born, its senses are subjected to even greater bombardment by visual, tactile, auditory and olfactory (chemical) stimuli. As it grows, the organism adapts and changes its behaviour in the light of these stimuli and its increasing experience. In the nest, many young birds spend a considerable amount of time flapping their wings as they build up their flight muscles. They may also take short, hopping flights on, and around, the nest. When they take their first real flight, though, they are able to fly immediately. However, their control on landing is often poor. A key environmental factor is the wind, especially its speed and direction. After a number of rather clumsy attempts, birds learn that for greater control it is best to land into the wind: similarly, it is best to take off into the wind for maximum lift. This simple example shows how an initially clumsy behaviour, that may be largely genetically determined, can be refined greatly by experience.

The singing behaviour of birds illustrates that environmental influ-ences can be significant in the development of song, as young birds learn to acquire the song pattern of adults. A number of species have been quite extensively investigated and one of these is the white-crowned sparrow

(*Zonotrichia leucophrys*). These are small American songbirds, of which only the males sing. Whilst they are chicks the males don't sing, but when they are a few months old they begin singing a form of song called the 'subsong'. If a young bird is isolated from others for the first 50 days of its life, after this time it will produce the subsong but not the typical adult male song. It seems that the young bird must hear normal male song during this 50-day period, which it memorises, and then bases its own song on this template or model. The period of 50 days is known as the **sensitive period** and is when young males are particularly sensitive to the songs of older males nearby. Song development in white-crowned sparrows therefore occurs in two stages – the early 50-day period when the song is learned and memorised, and a later stage, when the birds are about 6 months to a year old, when the males begin to sing and match their song to adult song they heard when they were a chick.

About 50 years ago the ethologist Niko Tinbergen observed that chicks of gull species, such as herring gulls (*Larus argentatus*) and lesser black-backed gulls (*Larus fuscus*), peck at a red spot on the parent's bill (figure 2.2), to encourage it to regurgitate food onto the ground for them to eat. Tinbergen used a series of cardboard cut-outs of a gull's head with different colour markings and measured how many pecks were delivered to each model head. He found that it was a model with a red spot near the tip of the beak that received most pecks. The red spot acts as a signal to the chick – when they peck at it, food is delivered. Tinbergen thought that the pecking at the red spot was instinctive. However, a researcher called Hailman later showed that, although pecking is instinctive, the chicks have to learn to peck

Figure 2.2   An adult herring gull. The red spot is on the lower mandible.

at the red spot. Initially the chicks peck anywhere on, or near, the parent's head but they soon learn that if they aim for the red spot they receive food quicker.

The above examples show how genetic and environmental influences interact in many behaviours. It is, therefore, probably more useful to consider most behaviours as neither wholly genetic nor wholly environmental, rather that they represent the two opposing ends of a continuum with particular pieces of behaviour somewhere on the continuum.

2.2   Instinctive behaviour

If there is a continuum of behaviours depending on the relative influences of genes and the environment, then at the genetic end there are behaviours that are under little, if any, environmental influence. These are termed **instinctive behaviours**. These behaviours are built into the organism at fertilisation – that is, they are largely under genetic control. This means that when the behaviour is needed by the animal, it is available in its complete form. It requires no practice and is normally released by a specific signal or stimulus. The stimulus produces a fixed response from the organism. Although instinctive behaviours can be seen in animals, this does not mean, as we shall see, that they cannot be modified in response to environmental factors.

Konrad Lorenz and Niko Tinbergen, the pioneers of animal behaviour, imagined that instinctive and learned behaviours were much more distinct than we now know is the case. This is shown in the four criteria that they suggested we use to allow us to identify an instinctive behaviour. These are that the behaviour:

- should be stereotyped;
- should be characteristic of the particular species (that is, be species-specific);
- should be seen in an animal that has been reared in isolation since birth;
- should appear fully developed in animals that have had no opportunity to practise the behaviour.

However, these elements are not as rigid as Lorenz and Tinbergen thought.

As far as instinctive behaviour is concerned, it is certainly advantageous to an organism to have the necessary neural circuits already pre-wired in its brain so that it can react appropriately when it needs to do so. These pre-wired circuits allow it to behave appropriately by reducing the time needed to process the stimuli, to decide how to react to the stimuli and to respond to them by engaging in the desired behaviour.

One of the pieces of behaviour identified by early researchers as instinctive was egg retrieval by greylag geese (*Anser anser*). Geese are ground-nesting birds and occasionally an egg may roll out of the nest. If this does occur, the goose stretches out its neck, puts its beak on the far side of the egg and, guiding it with its bill, rolls it back into the nest (figure 2.3) to continue the incubation. All greylag geese perform the act in the same stereotyped way. In fact, so stereotyped is this behaviour that, if a researcher close to the nest removes the egg that is being rolled, the goose continues to roll the non-existent egg! To humans, this seems strange indeed and irrational. It is as if the behaviour sequence was like a loop of film – the sequence is triggered and continues all the way through and only stops when the credits come up at the end of the film. Of course, in the wild other animals aren't removing eggs as they are being retrieved and so the behaviour is appropriate.

These forms of stereotyped behaviour were termed **fixed action patterns** by Lorenz – all individuals of a particular species perform the same behaviour in the same way in the same situation. Lorenz termed the stimulus that produced this sequence of behaviour a **sign stimulus** or **releaser,** since it 'releases' this chain of events. Other researchers have identified other sign stimuli – for example, the red spot on the parent gull's beak (figure 2.2) and the red belly of a male stickleback in the breeding season (section 1.4).

Figure 2.3 The egg retrieval behaviour shown by the greylag goose.

2.3   Attachment and imprinting

For most insect and many fish species very limited parental care is given to offspring. Any care that is given is usually from the female and concerns the selection of a suitable habitat to deposit her eggs. However, for most birds and mammals there is a considerable amount of care given after the young have been born. Indeed, the young and (usually) the mother often show **attachment** – that is a bonding together which normally keeps them in close proximity.

In the 1950s and 1960s a number of studies were carried out to demonstrate attachment between a parent and its offspring. A particularly well-known study was that by H. Harlow in the USA who investigated attachment shown by infant rhesus monkeys (*Macaca mulatta*). The infants were taken from their mothers shortly after birth and placed in individual cages. Each cage had two wire mesh cylinders, each large enough for a young monkey to climb on and cling to. A teat was fitted to one model which delivered milk to the infant, the other wire model had a towelling cover. Both models had heads and faces too (figure 2.4).

Before Harlow's experiment, most scientists believed that an infant monkey would spend most of its time on, or near to, the 'mother' supplying milk, since it supplied food. However, the experiments revealed that

Figure 2.4   The two artificial mothers used by Harlow in his experiments. The infant monkey spent most time clinging to the towelling mother.

although the infant monkey would visit the wire mother to feed, it would spend most of its time clinging to the towelling mother. If frightened, the infant would always run and cling to the towelling mother. Clearly, attachment cannot simply be explained by a mother feeding her infant. The young monkey needs the tactile comfort provided by the towelling material.

These are seen as drastic and even cruel experiments now, since isolating the young monkeys led to very disturbed behaviour in some of the monkeys. However, it led J. Bowlby, a British child psychologist, to believe that a similar process might occur in human babies and as a result changes were made in nursing and child-rearing practices in maternity wards in British hospitals.

Attachment is seen in young children and is evident if a child is separated from its mother. If distressed, a child will seek out and want to interact with its mother, often seeking tactile and verbal reassurance from the parent.

Some mammals and birds produce young that are well-developed at birth and soon become mobile and relatively independent – these are **precocial** young. Others, like human babies, are helpless at birth and entirely dependent on parental protection – these are **altricial** young. In a number of bird species with precocial young a special form of attachment called **imprinting** occurs. Imprinting is the process by which a young bird develops a **preference** for following its mother. Once established, the preference for following the mother is generally unfailing and irreversible.

Why should offspring benefit from following and being in close proximity to their parent? Animals grow up in a world with many potential predators with only one, or perhaps two, adult animals to protect it. Hence it is vital for the survival of the young to recognise and follow the parent and thus benefit from its protection. A parent, too, wants to recognise its own young and so protect its genetic investment in future generations. So both parent and offspring need to recognise each other.

Lorenz was the first researcher to study imprinting. Lorenz's studies were carried out on greylag geese. He, like others before him, had noticed that the young geese learn to follow their mother around within a few hours of hatching (figure 2.5). He believed that imprinting:

- was a fixed form of attachment and irreversible, so whatever the young imprint on they remained attached to and follow wherever it goes;
- took place within about 24 hours after hatching – this he termed the **critical period.** (Lorenz used this term to suggest that the following response had to be triggered by the appropriate stimuli at a particular time in the development of the chick. It was as if a developmental opportunity or window was opened up at this time and, if the necessary stimuli were available, imprinting occurred. However, this implies a fixed time available for imprinting whereas we now know this is not the case. Hence the term 'sensitive period' is preferred.)

Figure 2.5 A pair of greylag geese with their imprinted goslings close by.

Although the facility to imprint is presumably pre-wired into the neural circuitry of the gosling, learning is also involved. The bird is predisposed to learn to follow the first large moving object it sees. By carrying out a series of experiments, Lorenz found that goslings could learn to follow any large moving object, such as himself, watering cans, wellington boots and balls – the latter three being attached to rope and pulled along by the researcher.

Why would goslings want to follow a large moving object? Well, in nature, the object is almost always the mother who leads the young away from the nest where the tell-tale signs of broken eggshells suggest to aerial predators that easy pickings are available (see section 1.4). The mother also shows them where food is and offers protection from ground and aerial predators. It might happen occasionally in the wild that the mother is not at the nest during the sensitive period so there is a chance that the goslings might follow an inappropriate object. The chances of this happening are greatly reduced, however, as the imprinting process is not based solely on the goslings responding to visual stimuli alone. Other research with chickens has shown that in the egg, shortly before hatching, the chicks learn to recognise the mother's clucking sounds. Young birds actually like to follow the first large moving object they see that makes the sounds they heard whilst they were in the egg.

Some recent work involving domestic chicks (*Gallus gallus domesticus*) has shown that, if chicks are offered a visual stimulus to imprint on *plus* recorded sounds of maternal clucking, they are more likely to approach the stimulus object during the imprinting process. This might be because they pay greater attention to, or are more aroused by, both the visual and auditory stimuli, or perhaps because they learn to associate that the two signals occur together.

Since Lorenz's pioneering work, many other researchers have investigated aspects of this following response by chicks. Work by P.P.G. Bateson and others published in 1990, in a study using two different coloured shapes for the stimuli, showed that the preference for the imprinting object can be changed if the original stimulus is removed and replaced by another – in this study there were two different coloured shapes. However, they also found that if the chicks were isolated for a few days and then exposed to the first stimulus object again, a number of chicks switched back to that original stimulus. There is more flexibility in imprinting than was originally thought!

The majority of studies on imprinting have been carried out on bird species that have precocial young. What about young chicks that spend many days or weeks in a nest cup? Do they recognise their parents?

A study reported in 1993 shows that imprinting may also occur in altricial young. In this study blackbird nestlings (*Turdus merula*) were used. These stay in the nest for a number of days after they have opened their eyes. As a consequence, the nestlings have a while to adjust to, and recognise, their parents. In precocial species, imprinting occurs shortly after birth and is a very rapid process. Would the same be the case for blackbird nestlings? The study birds were placed in artificial nests when seven days old, just before they opened their eyes. They were fed on a diet of earthworms, fruit and commercial food, the feeding being done using surrogate parents – that is, either a stuffed male blackbird or a couple of small cardboard boxes glued together to mimic the body and head of a blackbird. The researcher found that a single feeding session was quite enough for a nestling to imprint on the parental object. After just a two-minute exposure period to the 'parent' that fed them, they would recognise them a second time, even if it was 24 hours later. Recognition was defined as the preference for begging from the familiar object (that is, the one that they had initially been fed by) rather than an unfamiliar object. So it seems that, in species with precocial young and some with altricial young, the identification of the parent is extremely rapid.

Although most of the research on imprinting has focused on birds, imprinting-like behaviour has also been noted in mammals, for example shrews (*Sorex araneus*). Young shrews need to recognise and follow their mother as she leads them from one area to another. In fact they form a 'caravan' as each young shrew grabs the fur of its sibling (or its mother) and they all move off, linked together as a line of animals. There is a sensitive period of 8–14 days during which the young shrews become imprinted on the smell of their mother.

The type of imprinting we have outlined above is termed **filial imprinting** because it describes the early and rapid process by which young birds and mammals recognise and follow their parent. There is another type of imprinting, called **sexual imprinting,** in which the early experiences of a young animal affects its choice of sexual partner later when it is mature. For example, a young male gosling learns to recognise its mother and its female sibs; later, as an adult, it will choose a different female goose for a mate. Thus there are long-term consequences of this type of imprinting. A gosling that has imprinted on a human or a watering can will experience considerable difficulty later in life when trying to find a sexual partner! However, there is some flexibility here too, as was shown in one study. This research used ducklings that were fed by a worker who always wore yellow rubber gloves when presenting them with their food. The ducklings imprinted on the gloves. When the males were sexually mature they were offered yellow rubber gloves and tried to court and mate with them. However, when they had the opportunity to court female ducks they switched allegiance!

Further insights into sexual imprinting came with experimental work on zebra and Bengalese finches (figure 2.6). These experiments involved cross-fostering, that is zebra finches were reared by Bengalese parents and vice versa. These studies showed that the zebra finch males tried to court Bengalese females and sang the song of Bengalese male finches, having modelled their song on that of their foster father. Similarly, Bengalese males tried to mate with zebra females. However, when a fostered male was housed with a female of its own species it did usually mate with her. However, when subsequently given a choice of potential female partners it preferred a female of its foster species.

Figure 2.6 Adult male zebra and Bengalese finches: the zebra finch is on the left and the Bengalese finch is on the right.

As with filial imprinting, there seems to be a sensitive period for sexual imprinting. It seems that it is not complete until the chick is several weeks old. Bateson has suggested that this is particularly advantageous since it is not until a bird is several weeks old that it develops its adult plumage. It is best, therefore, to delay sexual imprinting until all the other birds around it are adult. Bateson, who worked with Japanese quail (*Coturnix coturnix japonica*), showed that male birds prefer to mate with female partners who had slightly different feather markings to their mother. The fact that males chose mates with *fairly similar* plumage markings is important. 'Similar' would ensure they chose a female of the same species, 'fairly' would ensure they did not mate with their mother or their sister and so avoid the danger of inbreeding.

2.4 Play

Anyone who has looked after a puppy or a kitten is aware that they spend quite a considerable amount of time in play. Play seems to be an important part of the daily activity of young mammals and of some bird species, especially birds of prey. **Play** is an energetic activity, often involving two or more animals, during which the animals engage in sequences of behaviour which may be repeated a number of times. One of the features of play is that it often includes behaviour that is typically 'adult' but without the adult consequences – for example, one animal may bite the body of another, but the bite will rarely inflict any injury on their playmate. It seems, therefore, that play may not have an immediate adaptive purpose, but the benefits may be found later in life.

Young animals typically spend 10% or less of their time in play. However, this figure is slightly misleading since young animals spend a large amount of time resting and sleeping. So the amount of active time spent playing can be substantially higher than 10%. In a similar way, the total amount of energy a young animal expends in play may appear to be quite low, but much of the energy expenditure is given to their resting metabolic rate and to growth. If these 'fixed' amounts of energy are taken into consideration then the energetic costs of play can be quite high.

Attempts have been made to identify different types of play. Studies of cheetah cubs (*Acinonyx jubatus*) allowed one researcher to identify five types of play, see table 2.1. A study of young sable antelopes (*Hippotragus niger*) revealed seven patterns of behaviour in just one type of play – that is, social play (figure 2.7).

What might be the functions of play? Some of the possible functions include the following.

- **Play may help to develop the motor and cognitive skills of the animal**.

  The study of cheetah cubs described above suggests that locomotor

Table 2.1   Play types, their behaviour patterns and the recipient involved in the play of cheetah cubs.

Type of play	Behaviour patterns	Recipient
locomotor play	bounding gait rushing around	none
contact social play	patting biting kicking grasping	any family member
object play	patting biting kicking carrying	object
non-contact social play	stalking crouching chasing fleeing rearing up	any family member
exploration	sniffing	object

Definitions: bounding gait: slow run with stiff legs causing a rocking motion; rushing around: short sprint often including turns; pat: slap or touch with forepaw; bite: close jaws on animal or object; kick: strike with hindfeet; grasp: hold with forepaws or forelegs; carry: move with object in mouth; stalk: slow approach with body held low; crouch: stationary posture with body low and belly often on ground; chase: run after another animal; flee: run away from another animal; rear up: forelegs off the ground; sniff: place nose close to object.

skills are important to the young animals. The researcher found that locomotor play (for a description see table 2.1), in common with other types of play, declined over time – that is, from 2 months after birth, when the young emerge from their lair, until they are over 12 months old. Very young cubs engage in a lot of locomotor play which suggests that play may have an immediate benefit, rather than a delayed benefit, perhaps allowing cubs to escape from predators more easily.

- **Play may allow the animal to practise behaviour which will be important later in life, for example fighting and stalking prey**.
  Research on American kestrels (*Falco sparverius*) conducted in 1994 gave young birds the opportunity to play with twigs, corks, pine cones and model mice. The researchers recorded which 'toys' the young birds played with and for how long they did so. They found that the birds preferred mouse models to the other objects, presumably

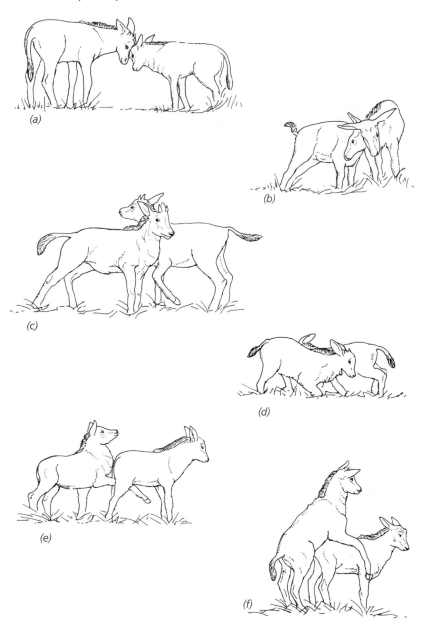

Figure 2.7 Social play patterns of infant sable antelope. These include: (*a*) butt – head or horn contact between two individuals (*b*) neck wrestle – individuals stand beside each other, facing the same direction, and push against each other's shoulders (*c*) shoulder – individuals stand beside each other, facing in the opposite directions, and push against each other's shoulders (*d*) whirl around – individuals stand beside each other, shoulder to hip, and as they push against each other they travel around in a circle (*e*) foreleg kick – one stands behind another and kicks with its foreleg between the hindlegs of the other (*f*) mount – one raises its forelegs over the back of another.

because they prefer to play with objects that closely resemble their usual prey.

- **Play may allow the young to familiarise themselves with novel items which may be useful to them as adults**.

  Studies of piglets (*Sus scrofa*) have shown that they have a high level of curiosity about novel objects. One study tested piglets with both familiar and novel objects placed in a side pen off the main enclosure. For each of the litters studied the piglets spent significantly more time in contact with the novel object than the familiar one. Another example is that of chimps playing with sticks whilst young which may help them later, when they are adults, as they probe mounds for termites.

- **Play may allow individual animals to gauge their physical condition against others of similar age**.

  The study of sable antelope calves mentioned above showed that calves initially play with their peers rather than with yearling or adult antelopes. Calves showed a preference for a few play partners of their own age but of either sex. Young sable antelopes seem to use play bouts to assess themselves and their partner with regard to their physical condition. Play therefore allows a calf to determine if it is growing faster or is more skilful and agile than its play partner. So play may offer an opportunity for self-assessment.

Recent research also suggests that other, perhaps more subtle, influences may affect aspects of play in young animals.

Although play has benefits, it has costs too. Occasionally, young primates fall and injure themselves when playing, while cheetah and lion cubs can injure a foot or leg if they don't spot a hole in the ground as they run and chase each other. The penalty can even be death. Southern sea lions (*Otaria byronia*) regularly catch South American fur seals (*Arctocephalus australis*). One researcher recorded 102 attacks on fur seal pups during a nine-month spell in 1988, of which 26 resulted in the death of the pup. Of these 26 deaths, 22 occurred while the pups engaged in play. The pups seemed to be caught because they were less vigilant whilst playing, not even noticing adult fur seals fleeing from the attacking sea lions.

THREE

Responding to the environment

3.1 Detecting stimuli

An animal's nervous system is constantly bombarded with information. Much of this information has no visible effect on the animal – think, for example, of how you can focus on just a fraction of the information you receive, paying no attention to the rest. In a conversation in a crowded room, for instance, you can listen to just one or perhaps two conversations while ignoring all the others. Interestingly, if someone within earshot, to whom you have not previously been paying attention, mentions your name you rapidly switch attention to what that person is saying. This shows that our brains are capable of monitoring more of the information in our environment than we might think.

The objects or events in an animal's environment to which it does respond are called **stimuli**. Most stimuli come from outside an organism, though organisms also respond to internal stimuli. Stimuli are detected using a range of sense organs.

- **Eyes** contain **photoreceptors** which detect **light** (electromagnetic radiation). Mammalian eyes differ in their sensitivity to light intensity, with nocturnal species being more sensitive to low light intensities than diurnal ones. However, mammals can only detect light with a wavelength between about 380 nm (blue) and 680 nm (red) whereas many insects can see in the ultraviolet. Sunlight vibrates predominantly in one plane, in other words it is polarised and some insects, including honey bees, are sensitive to the plane of polarisation of light.
- **Ears** detect **sounds** (air waves). Human ears can detect sounds with frequencies of between about 20 and 18000 Hz. Many mammals can detect higher frequencies than this (ultrasound). Bats and cetaceans use ultrasound in echo location, while the young of many small mammals emit ultrasonic squeaks if separated from their mothers.
- **Chemoreceptors** detect **tastes** and **smells** (the results of airborne or waterborne chemicals). Compared to many species, humans have relatively poor senses of smell (olfaction) and taste. Many other species can discriminate between a greater range of chemicals than we can or can detect chemicals at far lower concentrations. Sniffer dogs are

famous for this, while the males of certain moths can respond to just a handful of molecules released by females of the same species. These molecules are **pheromones,** airborne chemicals that carry messages, in this case signifying that the female is ready to mate.

- Some species, including bees and certain birds, can respond to variations in the Earth's **magnetic field** and use these variations for the purpose of navigation.
- A few fish generate **electric fields** and then use distortions in these fields to gather information about the position of potential prey.
- **Pressure receptors** in the external covering of an organism respond to **touch** and **pain.** Fish possess a **lateral line organ** – a row of pressure detectors along each flank – which allows them to detect mechanical vibrations in the water.
- **Thermoreceptors** respond to **heat** and **cold** and are found both on the surface of animals, where they respond to ambient temperatures, and inside animals, where they monitor the animal's own internal temperature.
- **Proprioceptors** detect changes within an animal's body and are involved in the detection of position and movement and so help maintain balance. In humans, proprioceptors are found within muscles and in the vestibular apparatus of our ears.

3.2 Processing information

Between an animal detecting a stimulus and behaving in some way is the stage where it processes the information. This processing is done by the **central nervous system**. In many respects, our understanding of how the central nervous system processes information is in its infancy, though the field of neurophysiology, which studies the workings of the nervous system, is advancing quickly. Just think of the stages that lie between your seeing the symbols on this page that make up writing and deciding whether to read any further. For a start, your brain has to translate the stream of information it receives from your eyes into meaningful communications, in this case communications written by the authors of this book about your nervous system. For this translation to be successful, you have to understand the words we have written.

In addition, you must feel **motivated** to read. If what we have written doesn't make sense, you will probably soon give up – your motivation to read further will wane. Equally, if what we have written is boring then, however much sense it makes, you will again probably soon give up. Assuming you are motivated to continue reading, your motivation can be described as external or internal. If you are dutifully reading this section simply because you need the information in it to write an essay you have been set, or because you hope it will eventually get you a better grade in some examination, your motivation is **external**. If, though, you actually find it quite interesting and are reading it because you find it intrinsically rewarding, your motivation is **internal**.

Many factors affect an animal's **motivation**. For example, at the simplest level the shorter an animal is of water the more thirsty it is and the more motivated it is to obtain water. The extent to which an animal is motivated to carry out a behaviour can be measured. For example, if laboratory rats have been trained to press a lever to obtain food, then a measure of how often they will press a lever before being rewarded with a food item is a measure of their hunger.

An animal's motivation is often affected by the level of certain of its hormones. Consider courtship and reproduction in the ring dove (figure 3.1). A male ring dove normally courts as soon as it is put with a female. Male courtship is triggered by a combination of the sight of a female ring dove (an external stimulus) and circulating levels of testosterone (a hormone produced by the male's testes). A female ring dove will not lay eggs unless a male is present. In fact, experiments show that the *sight* of a *courting* male separated from the female by a glass screen is much more effective than the *presence* of a *non-courting* male. The sight of a courting male causes a female ring dove to produce follicle stimulating hormone (FSH) and luteinising hormone (LH) from her pituitary.

Figure 3.1 A ring dove (*Streptopelia risoria*). The courtship of this species has been much studied.

FSH and LH stimulate the growth of the female's ovaries. As a result, her ovaries secrete oestrogen which causes her reproductive tract to grow. Rising levels of oestrogen also affect her behaviour. The female builds a nest with the male and the pair copulate. Incubation in the female is then triggered by yet another hormone, progesterone. At this time, the male too secretes progesterone. This acts antagonistically to testosterone and the male also becomes willing to sit on the eggs.

The final hormone to play a role is prolactin. This is produced in both the female and the male and leads to the production of a rich liquid from the birds' crops (in their throats) called 'crop milk'. From a functional point of view, crop milk – produced by pigeons and doves – is equivalent to the milk produced by female mammals. The parent birds use their crop milk to feed their young. Once the young birds fledge, prolactin levels fall in both adults. This leads to an increase in FSH and LH levels in the female and an increase in testosterone levels in the male, and the whole reproductive cycle may begin again.

By and large, an animal's behaviour is appropriate to the stimuli it receives. Occasionally, though, an animal may be equally motivated to perform two behaviours. For example, a territorial male who is approached by a courting female may be torn between repulsing her from his territory and engaging in courtship with her. Sometimes, in these situations, an apparently irrelevant activity occurs which is called a **displacement activity**. For example, our male, unable to decide (consciously or unconsciously) whether to attack or court may instead tidy up the boundary of his territory! Much the same thing can occur in humans. If you are momentarily stuck between two alternative courses of action, you may run your hand through your hair or laugh or scratch your neck. Interestingly, such behaviour carries information to whoever is in your presence. That person knows that you are unsure which course of action to take. The same is often the case in non-humans. A female bird faced with a courting male who suddenly preens or needlessly tidies up his territory is being given information that the male is unsure whether to attack or continue to court. Indeed, such displacement activities seem sometimes to have become incorporated, over evolutionary time, into the courtship routines of certain species.

3.3 Principles of communication

Many of the most important external stimuli that reach an animal come from other animals. It is quite difficult to provide a comprehensive definition of communication but the following are the key components of a communication:

- the **sender** – the individual who produces a signal;
- the **receiver** – the individual whose probability of behaving in a certain way is altered by the signal;

- the **signal** – the message that is sent;
- the **channel** – the medium through which the signal is transmitted (for example, visual or chemical);
- the **context** – the setting in which the signal is transmitted and received (such as courtship, or a fight);
- **noise** – background activity in the channel which is irrelevant to the signal being transmitted;
- the **code** – the complete range of possible signals.

It is easy to assume that communication involves co-operation between sender and receiver. Well, it may, but it is probably more realistic to think of sender and receiver as two individuals each with their own interests. To put it succinctly, natural selection acts on senders to send information that is to their advantage and natural selection acts on receivers to receive information that is to their advantage. This distinction is most easily appreciated if we think of communication between individuals belonging to different species. Consider a caterpillar that blends into its background through **crypsis** (being hidden) and a bird that feeds on such caterpillars and other insects (figure 3.2). The cryptic caterpillar is communicating 'I am part of a plant'. Clearly this is not an honest form of communication. (As is so often the case in animal behaviour, words like 'honest', as used here, are not used in the human sense where terms such as 'honest', 'deceive', 'decide' and 'selfish' suggest that some *conscious* thought occurs. Here, honest means 'reliable' and 'accurate'.) The cryptic caterpillar is trying to deceive the bird. **Deception** in communication also occurs between individuals of the same species. Think of a domestic cat about to fight. Such a cat may fluff up its fur, trying to make itself look larger than it really is. The same phenomenon is seen in many species.

Figure 3.2 An Australian flame robin eating a caterpillar. The caterpillar is camouflaged to look like the plants on which it feeds.

Of course, natural selection works on receivers too. Receivers have evolved to filter out misleading signals and focus on what they want to obtain from signals. For example, a female bird being displayed to by a male uses the information contained in his display to conclude how healthy and fit he is.

3.4 Territorial advertisement

Many animals communicate with one another in territorial encounters. A **territory** is a more or less exclusive area defended by an individual or group. Obviously, it is only worth an animal's while defending a territory if some benefit results. There are a number of benefits that follow from territoriality and these vary from species to species. A frequent function of territoriality is to enable the territory holder to attract individuals of the opposite sex for mating. Another frequent function is to defend an area rich in food.

Territories for reproduction

Examples of territories held for the purposes of reproduction are provided by song birds such as titmice, thrushes and finches. In great tits (*Parus major*), adult males set up territories in the spring. The males advertise their presence chiefly by song but also display (figure 3.3). (Females sing and display only rarely.) In an experiment to see whether song did indeed play a role in territorial advertisement, John Krebs and his colleagues removed territorial great tits from an area of mixed deciduous woodland near Oxford. One part of the woodland (the control) was left alone, another part of the woodland had a tape-recorder installed that played one type of great tit song, and a third part of the woodland had a tape-recorder which played a recording of several great tit songs. The experiment was carried out on three occasions. On each occasion, the control area was the first to be colonised by a new male and the area in which several great tit songs were played was the last.

Evidently at least one function of great tit song is to deter territorial intrusion. But why did the recordings of several great tit songs 'defend' a territory for longer than the recordings of one great tit song? Male great tits actually have song repertoires – that is, they sing more than one sort of song. Krebs suggests that a repertoire of songs may be more effective than a single song due to the *beau geste* effect. In the novel of that name, French Foreign Legion troops successfully defend a site by propping up their dead comrades at the battlements, thereby convincing the attackers that there are more French troops than is actually the case. In the same way, imagine a situation in which each territorial great tit male sings only one song, though there are slight differences between the songs that the various males sing. Now imagine that a single male starts to sing two or three different songs.

Perhaps he would be able to persuade his neighbours that two or three males were occupying the one territory. This might lead his neighbours to conclude that the territory in question is rather crowded and so not worth encroaching on. This, in turn, might allow the male with the several songs to spend less time singing and so save time and energy for other activities. Singing several songs would then be at a selective advantage and would be expected, provided it had a genetic component, to spread through the population. Interestingly, once it had spread through the population it would still pay each male to sing several songs as Krebs' experiment showed.

You may think this story rather a fanciful one. However, it does show how biologists try to find functional explanations for behaviours and work out how the behaviours might have evolved. If you can come up with a better explanation for the results of Krebs' experiment, please write and let us know!

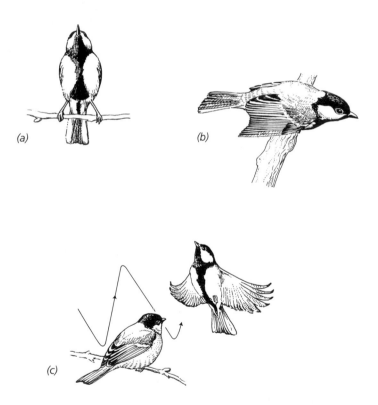

Figure 3.3 Displays in male territorial great tits. (*a*) The 'pointing display' which presents and exaggerates the width of the belly stripe. (*b*) The more threatening 'forward-threat display' in which the bill is aimed at the rival. (*c*) The 'pointing flight' which occurs only in escalated conflicts between males. If the recipient of a pointing flight retaliates, a fight may ensue.

Territories for food

An elegant example of a territory held for the benefits of food is provided by the winter feeding territories of the pied wagtail *(Motacilla alba)* on a meadow in the Thames Valley, as studied by Davies and Houston. In this study, some pied wagtails defended territories along a river while others fed in flocks around flooded pools nearby. The territory-owning birds exploited a renewing food supply, namely small insects which were washed up by the river onto the muddy banks. Owners typically walked a regular circuit around their territories, up along one bank and then back down the other side again (figure 3.4*a*). This behaviour generally makes it unprofitable for intruders to land. The most profitable place for an intruder to land would be a short distance ahead of the owner, as there the prey has had the most time to accumulate. However, if an intruder landed there, it would be easily spotted by the territory owner. If, on the other hand, the intruder landed a short distance behind the owner, it would be feeding on a stretch only recently depleted of food. Measurements showed that on those occasions when they did manage to sneak onto a territory undetected, intruders often fed at an unprofitable rate precisely because they were feeding on recently depleted land.

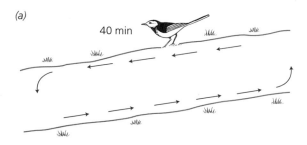

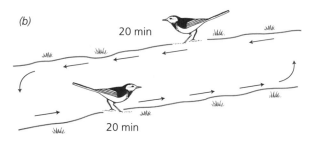

Figure 3.4 (*a*) Pied wagtails exploit their winter feeding territories systematically. At Davies and Houston's study site, a complete circuit took abut 40 minutes. (*b*) When two pied wagtails share a territory, each walks approximately half a circuit behind the other and so obtains about 20 minutes' worth of new food.

This may explain why intruders in pied wagtails are usually very noisy when flying over a territory, loudly calling 'chisick'. When this happens, the owner replies 'chee-wee' and the intruder usually flies off. Perhaps the intruder's noisy calls are an enquiry as to whether the owner is present. If the territorial bird is present, and calls, it pays the intruder to move on. This seems to be an example of **honest communication**. It benefits both the intruder and the territory owner to communicate in this way.

By way of contrast, there are a number of species of frogs and toads where smaller males try to hide in the mating territories of larger males. These smaller males do not engage in honest communication. Instead of croaking, as is typical of a territorial male frog, they keep very quiet and generally remain hidden. You can probably work out why this is so (their behaviour is discussed in more detail in section 6.3).

Returning to the pied wagtails, territory holders keep the same length of territory, a circuit of around 600 m, throughout the whole winter, despite changes in the amount of food on the territory. One might have expected the size of the territory to vary according to the amount of food on it. Such variation is known in certain hummingbirds, which very precisely alter their territory sizes according to the number of flowers that are flowering and the volume of nectar they produce. Perhaps continuous adjustments in territory size in pied wagtails would be too costly. At the start of winter, neighbours spend a lot of time in boundary disputes. Once settled, these boundaries are respected and maintained simply by short displays at intervals throughout the day.

However, although the territory size does not change in response to variations in food supply, the behaviour of the territory owners to intruders does. Four sorts of behaviour are found depending on food abundance.

- When food is scarce on the territories, the owners leave and feed elsewhere but keep returning at intervals to announce ownership and evict intruders. This suggests that territorial defence is a long-term investment and that a territory is worth retaining even through periods of low prey abundance.
- At intermediate levels of food abundance, owners spend all day on their territories and evict all intruders, as shown in figure 3.4a.
- At higher levels of food abundance, an owner shares its territory with another individual, a **satellite,** usually a first-winter juvenile or an adult female. Sharing the territory brings costs to the owner because the satellite obviously reduces the amount of food available to the territory owner. However, sharing brings benefits too, as the satellite helps with the defence against other birds. Presumably a single bird simply could not profitably defend an entire territory against all intruders at these high food levels. The original territory holder and the satellite minimise feeding interference by feeding half a circuit behind one another (figure 3.4b).

- At very high levels of food abundance, for example when there is a sudden emergence of insects in the spring, owners abandon all attempts at territorial defence and the birds feed in flocks.

Do humans have territories?

Whether or not humans are territorial is debatable. However, the zoologist Desmond Morris has argued that humans have three levels of territories: societal, family and personal.

- **Societal territory** – Humans evolved as group-living animals in units of probably only a few dozen individuals at most. As these bands grew into tribes, and tribes into nations, war chants became bugle calls and then national anthems, while territorial boundary lines hardened into fixed borders between countries. Patriotism, however, is rarely enough to satisfy us. Rare is the person who does not also belong to a small, more personal sub-group of society, for example a teenage clique or specialist association focused around some hobby or other activity, such as ballroom dancing or joy-riding. Some groups advertise themselves through badges, headquarters, slogans and other displays of group identity.
- **Family territory** – A family territory is typically centred around a home. Its boundaries are generally conspicuously displayed by markers such as walls or fences. Visitors who enter the family territory put themselves in a position where they may need to seek permission to do things, like sitting down, which they would consider a right in a neutral or their own environment. Many families are unable even to visit the seaside or enjoy a picnic without setting up a temporary territory, advertised by towels or rugs.
- **Personal space** – Each of us carries with us, wherever we go, a 'portable' territory known as our personal space. If people move inside this space, unless they are close (note the word!) family or friends, we feel threatened. Jammed into a lift with strangers, we are forced temporarily to abandon our personal space but behave as if there was no-one else there. We remain silent, adopt an expressionless set of facial features and typically avoid eye contact. Cultures differ in the size of people's personal spaces. In England, many people talk at roughly fingertip distance. In Mediterranean countries, the comfortable distance may be only half that. At international gatherings, threatened English people sometimes appear pinned to the wall by rejected Mediterraneans.

  As with societal and family territories, personal spaces may be advertised even if the territory holder is absent. If you place a jacket over the back of a chair in a library and then retreat, the chair will remain unoccupied for far longer than if you leave the chair unmarked.

Costs and benefits of territoriality

So, if territories have so many benefits, why don't all species have them? We have already touched on the answer in our consideration of the pied wagtail feeding territories. Territories are only defended when it makes sense to do so. Suppose that a territory is defended for the energy it provides for its territory owner. Imagine that, whatever the size of the territory, the energetic cost of defending is always greater than the energetic benefit of having the territory, as is illustrated in figure 3.5a. In this case, it is simply not worth being territorial. Imagine now that the benefit of holding a territory does exceed the cost, as in figure 3.5b. Now territoriality becomes worthwhile. Note that there may well be an optimal territory size, symbolised by T_{opt} in figure 3.5b. The optimal territory size is that at which the *difference* between the benefit and the cost – if you like, the pay-off from having a territory – is at a maximum.

3.5   Communication to resolve conflicts

Communication can serve many functions. One important one is to reduce the chance of an individual hurting itself in a conflict. In section 6.3 we shall look at how male red deer usually roar at each other before fighting.

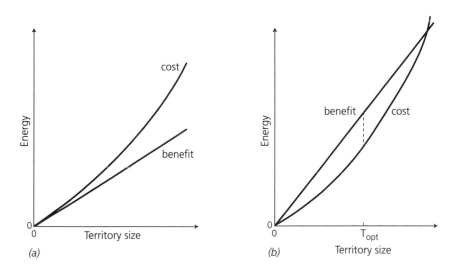

Figure 3.5   The energetics of territoriality. (*a*) The cost of holding the territory always exceeds the benefit, whatever the territory size. (*b*) For a range of territory sizes, the benefit of holding the territory exceeds the cost. T_{opt} indicates the optimal territory size, namely that at which the difference between the benefit and the cost is a maximum.

Fighting is dangerous and roaring enables two males to assess each other's physical condition before deciding whether to fight. Roaring is an example of an honest communication in which **ritualisation** occurs – fighting is preceded by a characteristic set of behaviours in which first roaring and then parallel walking, where the two individuals walk along side by side, occurs.

In many social species, individuals in a group arrange themselves in a **dominance hierarchy**. Sometimes these dominance hierarchies are known as **peck orders**, as they were first described in domestic hens which, as you might expect, use pecks to establish and maintain the hierarchy. The individuals in a dominance hierarchy can be ranked from top to bottom. Often these hierarchies are linear in the sense that if animal X is dominant to animal Y and animal Y is dominant to animal Z, then animal X will also be dominant to animal Z. In certain species, for example wolves and African wild hunting dogs, there are separate hierarchies for males and females.

In a species with dominance hierarchies, social interactions between a pair of animals are between a **dominant** individual and a **subordinate**. The benefits of being a dominant individual can be many. You may get access to better food or to high-ranking individuals of the opposite sex for mating purposes. Many studies have shown that dominant individuals leave more offspring than subordinate ones. Indeed, in certain species subordinate males never reproduce at all. Further, dominance is inherited in the sense that, in many species, the offspring of dominant mothers grow faster, are healthier and themselves become dominant individuals.

Once a dominance hierarchy has been established, little or no behaviour is needed to maintain it for considerable periods of time. Each individual knows its place. Dominant individuals show gestures of dominance and subordinates show gestures of **appeasement**. Here are a few examples.

- Dominant rhesus macaques have a leisurely stroll with head held high and tail erect.
- Subordinate African wild hunting dogs lie on their backs exposing their underbellies to the dominant individual which may lunge at them and bare its teeth. Subordinates also wag their tail between their legs and pull back the corners of their mouth in a wide 'grin'.
- Chimpanzees show a wide range of behaviours depending on the social context and on whether they are dominant or submissive. A subordinate is likely to 'pant-grunt' when approaching a more dominant individual or after being threatened or attacked. Pant-grunts consist of a series of rapid grunts, connected by audible intakes of breath. The jaws are often partly open and the teeth are normally concealed by the lips. If the dominant chimpanzee behaves at all aggressively, pant-grunting in the subordinate quickly gives way to squeaking or screaming and the subordinate 'grins' with a facial expression close to that of the rhesus macaque in figure 1.1.

3.6 The bee language controversy

More than 2000 years ago Aristotle found that, although a source of food placed near a honey bee (*Apis mellifera*) hive might remain undiscovered for hours or even days, once a single bee had located it, many new bees soon appeared. In a series of classic experiments von Frisch investigated how such behaviour occurred.

When worker bees collect high-quality food within about 80 m of the hive, they execute a **round dance** on their return to the hive (figure 3.6). Because it is normally dark in a colony (whether the colony is a natural one or lives in a hive provided for it), other workers do not watch the dance. Instead, they follow the dancing bee about on the comb and sense the vibrations she produces. The dance contains information about the quality of the food source. The better the food source, the more often the dancing bee changes the direction of her dance. However, the round dance contains no information about the direction of the food source, and the only information contained about the distance to that food source is that it lies within about 80 m of the hive.

The more often that a bee dancing the round dance changes her direction, the more likely following bees are to fly from the hive in search of the food source. They are helped in their search by the fact that, in addition to knowing that the food lies within some 80 m, they will have picked up smells adhering to the body of the dancing bee, and may have tasted regurgitated nectar.

If a bee finds a rich food source more than about 80 m from the hive, it performs a **waggle dance** on its return (figure 3.7). This conveys more precise information about the distance to the food source and also conveys information about its direction. The information about the distance covers

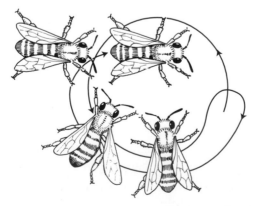

Figure 3.6 The round dance of the honey bee. The dancing bee is at the top of the drawing and the path of her dance is indicated by the arrows. The other three bees are following her.

the range from 80 to approximately 1000 m and is provided in three different ways:

- the speed with which the bee runs through a complete dance circuit;
- the number of abdominal waggles given during the central portion of the dance;
- the frequency with which sound bursts are produced while dancing.

It would, in fact, be more precise to say that the waggle dance provides information about the effort that needs to be expended on the flight to the food source. If the bee has to fly upwind or up a hill to get to the flowers, the recruiting worker produces fewer waggles, performs its dance circuits in a more leisurely way and gives a lower frequency of sound bursts.

The waggle dance also carries information about the direction of the food source (figure 3.7). The angle of the central portion of the dance clockwise from the vertical is the same as the angle that the food source makes clockwise from the Sun. Interestingly, von Frisch found that bees could even give and receive such information on cloudy days. This is because, as we mentioned above (page 28), their eyes are sensitive to polarised light. Polarised light is available even on cloudy days and from it, the direction of the Sun can be deduced.

It might be thought that there would be a difficulty with the way in which the bees communicate the direction of the food source because the

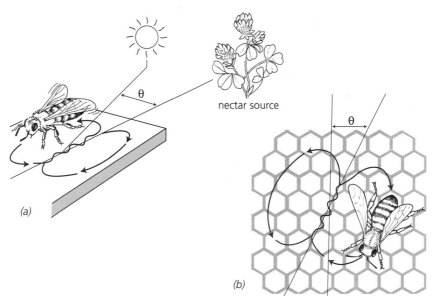

nectar source

(a)

(b)

Figure 3.7 The waggle dance of the honey bee. (a) The way in which the dance indicates direction is clearest when the dance is occasionally performed outside the hive on a horizontal surface. Here the bee runs directly at the food site. (b) On a vertical comb, inside the dark hive, the dance is oriented with respect to gravity. The angle that the central portion of the dance makes clockwise from the vertical equals the angle of the food source clockwise from the Sun.

Sun's position relative to the food source changes during the day. Accordingly, one might expect a bee's information to become more inaccurate with the passage of time, if it spends more than, say, half an hour in the hive dancing before re-emerging and getting new bearings. Remarkably, if dancing bees remain in the hive for any length of time, they perform their dances at a different angle to the vertical, correcting for the apparent movement of the Sun.

Objections to von Frisch's theory

Despite early experimental support for the bee language hypothesis of von Frisch, later experiments carried out by Wenner and his colleagues suggested that recruits might locate food sources by olfaction (smell) alone. Wenner did not dispute von Frisch's finding that the waggle dance contains information about the distance and direction of food but questioned whether recruits make use of this information. Wenner pointed to several instances where the behaviour of animals contains information decipherable to humans but not used by the animals concerned. For example, the movement of flies having exhausted a food source conveys information to a human observer about the type and concentration of the food source but other flies do not appear to use this information.

Wenner's criticisms stimulated others to attempt new experiments to test von Frisch's theory further. One series of experiments providing impressive support for von Frisch's theory was performed by Gould. Gould made use of the fact that if a small point source of light is provided at the side of a vertical comb, the bees treat this as if it was the Sun. This means that when they dance, bees treat this point source just as the bees dancing at the hive entrance on a horizontal surface treat the Sun (figure 3.7*a*). They angle the central portion of their waggle dance directly at the food source. Having provided such a light source, Gould then blacked out the three ocelli (simple eyes) on the dorsal surface of the head of some workers returning to the hive from rich food sources. Such bees could not detect the point source of light and so oriented their waggle dances on the vertical comb with respect to gravity. The other bees, however, interpreted the dances as if the light source was the Sun. Gould then moved the light on the comb a certain number of degrees every 30 minutes. As predicted by von Frisch's theory, the recruited bees' flights were shifted by the same number of degrees.

Despite Gould's experiments, some controversy about the extent to which bees use information from the waggle dance continued. Eventually, bee researchers succeeded in making tiny model bees which could be moved over combs as if the bees were performing waggle dances. The model bees have provided further evidence in support of von Frisch's and Gould's findings. They are also helping to unravel the precise way in which the waggle dance conveys information about the distance of the food source.

3.7 Language

At first sight, communication in humans might be thought to be fundamentally distinct from communication in other species. However, it should be borne in mind that much of the information that people exchange is by **nonverbal communication** as it is in other animals. For example, try moving your eyebrows to indicate greeting, suggest approval, invite flirtation, register indignation, express curiosity and admit surprise. Non-verbal communication can be conscious or unconscious. Both men and women register approval of objects they see by unconsciously enlarging their pupils. Accordingly, when heterosexuals are given a choice of photographs of people of the opposite sex whose pupils have been touched up so as to look larger versus unadulterated photographs, they usually prefer the photographs with the larger pupils.

Communication using 'true' language has traditionally provided a clear-cut separation of humans from other animals. By 'true' language is meant both the use of **symbols** (such as written words or spoken sounds) for abstract ideas (for example, 'predator') and an understanding of **syntax** – meaning that symbols convey different messages depending on their relative positions (for example 'woman eats plant' conveys a different message from 'plant eats woman').

There is no doubt that some species other than ourselves use symbols for communication. For example, vervet monkeys have three different types of alarm call. One is given when a leopard is sighted. The other monkeys respond by running into trees. A second is given when an eagle flies overhead. The other monkeys look up into the sky or hide in bushes. The third type of alarm call is given when snakes are spotted. On hearing it, the other monkeys look around at ground level. Playbacks of tape-recordings show that the three different alarm calls do genuinely carry these distinct messages as the monkeys behave to playbacks as they do to real monkeys making the calls.

Whether any non-humans really understand syntax is debatable. Several researchers have spent many years training captive chimpanzees, gorillas, parrots, pigs and other animals to learn words (in the case of parrots) or sign language (in the case of apes) or other symbols (in the case of pigs and some apes). For example, the Gardners used the American Sign Language for the deaf to teach a chimpanzee, Washoe, over 100 words. Another chimp, Sarah, was taught by Premack to use plastic blocks as word symbols. While many of these animals show an impressive richness of symbolic communication, it is clear that none of them comes close to the sophistication of human spoken language. Nor should we forget that humans have written as well as spoken languages. Writing has proved invaluable as a means of transmitting culture over the generations. There is no doubt that our communicative powers lie at the heart of human society.

FOUR

Learning

On a cold winter's day a blue tit (*Parus caeruleus*) alights on a branch of a tree in a garden. The owner has thoughtfully tied a peanut to a piece of string and fastened the string to the branch. Peanuts are a favourite food item for blue tits and the bird hops along the branch to the string. The tit takes hold of the string in its beak and pulls it up. It then uses one of its feet to hold the string tight and bends over again to grasp the string in its beak. By repeating this action several times the bird hauls up the peanut onto the branch and, using its feet to keep the string still, pecks at the nut (figure 4.1).

How was the bird able to link all these actions together and complete this complicated piece of behaviour? Did it rely on a set of instructions in its brain, instructions that were there at fertilisation? Did it perhaps see other birds behave in this way?

The answer is that the blue tit is probably not programmed on hatching to manipulate string in this way – after all, only a few gardens have such arrangements. The bird's behaviour is likely to be the result of learning. This does not mean, however, that all the elements of the tit's behaviour must have been learned. Tits, for example, regularly use their feet to hold food, pecking at the food with their beak. Holding items in their feet could have a genetic component.

Figure 4.1   Blue tit using its feet and beak to pull up a peanut tied to the end of a piece of string.

We all think we know what learning is. (Hopefully, it is what you are doing at the moment!) Learning is something that is not just restricted to your school or college environment. Nor is learning always 'correct' – for example humans can learn bad driving habits. Humans learn all the time, even when in situations where it isn't obvious, such as chatting with friends around a lunch table. Animals are also learning, though the amount of behaviour that is learned depends on the species. So the ability of mammals to learn is greater than that of insects, although some insects are very good at learning, for example navigating back to the nest after foraging for food. Further, learning should not be seen as the opposite of instinct, they can complement each other. We have already noted, for example, how gull chicks peck instinctively but that learning refines this behaviour so that when they peck at the red spot on the parent's bill they receive food quicker (page 16).

Learning can be defined as a relatively permanent change in an organism's behaviour which results from its experience. For many animals, much of their behaviour is learned. They rely on learning to find food, navigate, search for a mate, build bonds or relationships with others and recognise predators. How are such behaviours learned and how can scientists determine if learning has taken place?

A feature of learning is that we cannot see it occurring. It is a process and so we infer that it is happening, or has happened, from a change in the behaviour of the organism. What scientists measure is what is, or has been, learned. In the case of the blue tit, the scientists would observe and record the bird's behaviour over time. At some time in the past, when the bird had its first experience with the string and peanut, it would not have gone through the fairly smooth sequence of events described above. Repeating and refining its behaviour over time allows a bird to handle the nut quickly and efficiently. How can a bird manage to do this? The answer is by using one, or more, of a number of forms of learning.

There are a number of forms of learning that animals can use: these are habituation, classical conditioning, operant conditioning, trial-and-error learning, latent learning, insight learning and observational learning. (We have already referred to imprinting, pages 20–24, which is a form of learning that occurs very early in an animal's life.)

4.1 Habituation

Habituation is the decline in response to a specific stimulus over time, when that stimulus is repeatedly presented to the organism. It is the simplest form of learning as it is limited to the repeated exposure of *one* specific stimulus.

A sudden noise invariably startles animals. This is an adaptive reaction – the animal looks around to locate the source and takes the appropriate action. When the stimulus is repeated, and if the sudden noise has not proved dangerous, the animal is again startled, but to a lesser extent. If the stimulus is repeatedly presented, the animal's reaction lessens and eventually it ignores the stimulus altogether: habituation has occurred. If a different stimulus is presented, the animal will respond vigorously showing that the habituation is not just exhaustion and that the decline in response is stimulus specific.

What is the adaptive function of habituation? It is important for all animals to respond rapidly, and appropriately, to a new stimulus. Habituation allows the animal to respond to a new stimulus, which might be beneficial or harmful, but to ignore what it has learned to be a neutral stimulus. After a few minutes without stimulation the effect of habituation lessens and the organism will again respond to the stimulus.

In the past, some scientists have recorded alarm calls that animals make if a predator is near, using an audio tape-recorder. These recordings are then taken to similar habitats and played back when no predators are around so that the scientists can see how the prey animals respond. Can you suggest one danger for the animals if this experimental procedure is repeated for a number of days?

4.2 Classical conditioning

When habituated, animals learn not to respond to a stimulus. The stimulus is not linked, or associated, with any other event. In other forms of learning the organism associates one event with another. Dogs often associate being taken for a walk with the fact that their owner picks up a lead, or says 'Walkies!'. We can, therefore, divide learning into associative and non-associative learning: habituation and imprinting are types of non-associative learning; classical conditioning, operant conditioning and trial-and-error learning are types of associative learning.

The first experimental studies of associations between two, or more, stimuli were carried out by a Russian physiologist, Ivan Pavlov, at the beginning of the twentieth century: this form of learning is called **classical conditioning**. Pavlov was initially interested in the salivation response of dogs to the presentation of food rather than the behaviour of the dog. Salivation is a reflex action, an innate response to the appearance of food and occurs in all dogs, and in humans too. During Pavlov's experiments, a dog was kept in a harness and a small glass tube in its cheek collected the drops of saliva (figure 4.2). Pavlov noticed that the dog started to salivate before it saw the food. Its salivation was triggered when it saw one of Pavlov's assistants approaching it with a bucket inside which was the food. Dogs would even salivate if they heard the food being prepared!

Pavlov decided to study this response in a more controlled way and for this he used a bell. The bell rang just before food was presented and he observed the dog's behaviour and collected the saliva produced. Pavlov

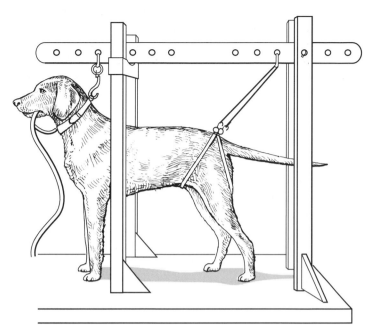

Figure 4.2 Apparatus used by Pavlov in his experiments.

carried out a whole series of these experiments, at the end of which he had identified the essential components of classical conditioning. He used specific terms to describe the elements in the learning process.

He called the food the **unconditioned stimulus** (UCS), the spontaneous release of saliva the **unconditioned response** (UCR), the bell the **conditioned stimulus** (CS) and the saliva produced in response to the bell the **conditioned response** (CR). The result of pairing the bell and the food is that the dog learns to expect food when the bell sounds. The dog has been trained to salivate to the sound of the bell, not something that dogs do naturally. The learning process can be summarised diagrammatically, as in table 4.1.

Table 4.1 Stages in the conditioning process.

Pre-experiment	bell	→	no saliva
Learning stages (a) Pre-training	food (UCS)	→	salivation (UCR)
(b) Training	bell + food (CS)　(UCS) [This stage is repeated a number of times]	→	salivation (UCR)
(c) Post-training (conditioning)	bell (CS)	→	salivation (CR)

What does the dog learn? The dog learns to associate two events, here the bell and the food. So it can predict the likelihood of an event occurring. Pavlov carried out many experiments and identified a number of features of conditioning. During the training stage the dog receives food when the bell sounds. If food is presented each time, the learning is reinforced or strengthened. The strengthening of the CR by repeatedly linking the CS and the UCS is termed **reinforcement**. However, if the food is not given after the dog has learned to link the bell and the food, the dog stops salivating, that is **extinction** occurs. The dog stops salivating because food is no longer given when the bell sounds. If, after extinction has occurred, the dog is left for a few days and then the bell sounds again, it typically salivates; this is termed **spontaneous recovery**. Spontaneous recovery is, however, short-lived, unless the UCS follows. Pavlov also found that if, after the training programme is established, a bell of a different tone rings then the dog salivates: the dog shows **generalisation**, that is it makes the same response to a different, but not too dissimilar, stimulus. The conditioning process can be used to get the dog to show **discrimination**, that is it salivates to only one of two bells. The researcher simply selects one bell and only gives food when that rings, but not the other. Pavlov also found that **higher-order conditioning** is possible. If, after training the dog to salivate to a bell, a buzzer sounds just before the bell, the dog salivates to the buzzer: this is because the dog learns to predict that the bell follows the buzzer and this means food is on the way.

A limitation of classical conditioning is that the learning is linked to reflexes or very strong emotions. Much human and animal learning needs to be more flexible and operant conditioning is more useful for this reason.

4.3   Operant conditioning

In **operant conditioning** the animal has to learn to perform an operation in order to receive a reward. The operation it is required to do can be selected by the researcher. If a dog is in a room with an experimenter the dog can behave in a number of ways: it might bark, scratch at the door, turn around, sit on the floor, wag its tail, etc. The researcher can select from these possible behaviours *one* which she/he wishes the dog to repeat. If this behaviour is rewarded with food the dog is likely to repeat the behaviour, until it is full and wants no more food.

These ideas developed from research carried out by E.L. Thorndike in the USA in the first couple of decades of the twentieth century. He put cats into cages (called puzzle boxes) and observed their behaviour as they tried to get out and reach a food item that they could see placed outside the cage. Inside the box was a loop of string and if it was pulled the door opened. When a cat is first put in this box it will explore, and eventually it will pull the string and escape. On being returned to the box the cat will escape again. Thorndike plotted the time taken to escape and found that it declined slowly. By the end of the training sessions the cat would pull the string as soon as it was put into the box (figure 4.3). In this experiment the cat uses **trial-and-error learning** to escape. There is a gradual improvement in performance over time; there is no sudden insight into how to escape.

Twenty years or so later another American, B.F. Skinner, developed these ideas. He devised a special piece of apparatus to investigate learning in rats and pigeons (the apparatus is called a Skinner box). The box is a simple piece of apparatus consisting of a box with a gridded floor, a light, a lever and a food tray (figure 4.4). When a rat is put in the box it explores and eventually touches the lever. This releases a food pellet from a hopper outside the box and it falls into a tray where it can be eaten by the rat.

Subsequently the rat learns to press the lever to obtain food, so it has learned to predict that pressing the lever is associated with food. The food *reinforces* the rat's response, making it more likely to repeat the act. This form of learning is called operant conditioning because the animal has to operate on its environment to obtain the reward: the rat gets the food through trial-and-error learning.

An important finding from Skinner's experiments is that the rat does not always have to be rewarded with food in order to continue to press the lever. If the rat is rewarded with a food pellet after each press, this is termed **continuous reinforcement**. Skinner identified four other types, called **schedules of reinforcement**. These are:

- **fixed interval** – the food appears after a fixed time interval, say 30 seconds, provided the lever is pressed at least once;
- **variable interval** – the food is given after different time intervals, say 23 seconds, then 37 seconds;
- **fixed ratio** – the food is given after, say, every tenth press of the lever;
- **variable ratio** – the food is given after a varying number of lever pressings, say the sixth then the fourteenth.

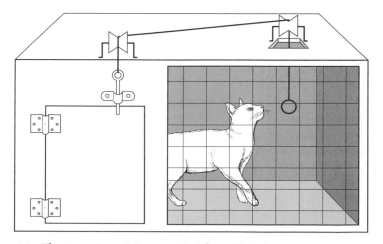

Figure 4.3 The type of puzzle box used by Thorndike in his experiments.

Not surprisingly, the rat continues to press the lever much longer with a variable ratio schedule because it cannot predict when the reward will come and carries on pressing in the hope of receiving food eventually. However, extinction does occur if the behaviour is not reinforced. By selecting the appropriate schedule of reinforcement, the experimenter can shape the behaviour of the organism. This technique is the basis of teaching tricks to circus animals, for example teaching seals to play horns and dogs to ride bicycles.

Operant conditioning, therefore, usually describes associative learning in a very unnatural, but highly controlled, environment. It has been popular with experimental psychologists who want to establish the rules by which animals learn but is criticised by ethologists because it ignores how the learning of a species may be adapted to its environment.

The food rewards the animal for showing the desired behaviour, pressing the lever. The behaviour has been **positively reinforced**, that is reward makes it more likely to be repeated. Positive reinforcement is used by parents to shape the behaviour of their children. When a child is learning to read, the parent may praise them when they identify a new word correctly. This will encourage the child to try to identify other new words. Humans are prepared to work for reinforcers that are not food or water: money and social approval can also be powerful reinforcers.

Figure 4.4 A rat in a modified Skinner box being offered a choice between two stimuli.

An alternative is **negative reinforcement**. When this is used an unpleasant effect, such as a mild electric shock, stops when the organism carries out the desired behaviour: all other responses result in a shock. Negative reinforcement is not the same as punishment. When **punishment** is given the organism receives a shock for showing behaviour that is not desired. The problem with punishment is that the animal does not recognise what it is expected to do, it only knows what *not* to do.

4.4 Latent learning

Latent means 'hidden': how can learning be hidden? In both classical and operant conditioning, reinforcers are crucial: without reinforcement, learning does not take place. However, in the 1930s E.C. Tolman, working with a group of scientists in the USA, carried out a series of experiments which showed that learning without reward was possible: the type of learning was called **latent learning**. This is learning that occurs but is not evident until the organism is in a situation in which the information it has learned can be used. It seems the organism's brain takes in and processes information in a particular situation and this information can be retrieved later when it helps the animal to solve a problem.

Tolman demonstrated this with rats in a maze (figure 4.5). The maze had two arms and in each box was food. The boxes could only be seen when a rat turned the corner at point X or Y. One box, box A, was painted white; the other, box B, was painted black. The same type and quantity of food was placed in each box. After a number of trials Tolman found that the rats were equally likely to go into each box. Subsequently, each rat was placed inside each box when the boxes were no longer attached to the maze. In box A they found food, in B they received an electric shock. After these experiences the rats were allowed to run in the maze under the initial conditions, namely, food in each box. Tolman found that the rats avoided box B. The animals were able to combine the experience from the earlier sessions and utilise it

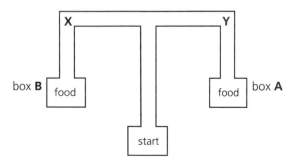

Figure 4.5   The type of maze used by Tolman in his experiments.

when it was appropriate, here to avoid an electric shock. Tolman claimed that the rats had built up a **mental** (or **cognitive**) **map** of the maze and they were able to draw on this information when it was useful to them.

We should not be surprised that animals show they can learn in the absence of reinforcement. For example, many species of social insects navigate around their environment with the help of learned landmarks. Organisms pick up visual and chemical signals, for example, as they move around their territory. These signals are stored and processed in the brain and can be used when appropriate, for example to bolt for the den when an aerial predator appears.

4.5   Insight learning

In the puzzle box the cat finds how to escape using trial-and-error learning. Does this mean the cat behaves intelligently? In fact, there was nothing else inside the box except the string. If the cat doesn't interact with the string it doesn't escape. Thorndike found that the cats gradually improved their performance but that learning was slow. Can animals show a flash of inspiration in problem-solving? Why didn't the cats suddenly 'see' how to escape? Why did they not use insight? They might have done had they been able to see the string, levers and pulleys that opened the door but they were hidden from view.

If all the elements of a problem are in view can animals solve problems using insight? W. Köhler believed that he demonstrated **insight learning** in chimpanzees (*Pan troglodytes*). In his experiments the chimps were put in an enclosure inside which was a variety of objects. In one experiment he placed a banana just outside the enclosure, but beyond the reach of the chimp. There were sticks in the enclosure and Köhler found the chimps put one stick inside another to make a rake which they used to bring the banana within reach. Köhler believed that, so long as all the elements of the solution were available and it was a problem within their capabilities, chimps could solve such problems using insight.

However, since sticks are common objects in the wild – in fact chimps use them to probe termite mounds to get and eat termites – their behaviour can be explained as trial-and-error learning. Further, since the enclosure and objects were familiar to the chimps they could use latent learning to solve the problem.

It is not easy to demonstrate conclusively that insight learning occurs in animals. One way might be to put the animal in a situation for which a natural solution could not have prepared it. Insight was apparently shown by a female macaque monkey that lived on a Japanese island. The island was used by scientists for research. Occasionally the scientists would put out food for the monkeys and one day sweet potatoes were dumped on the beach. Inevitably, the potatoes got covered with sand, especially as the monkeys competed for them. One female was seen to take her potato to the water and wash off the sand. This behaviour was copied by others. Some time later wheat was put out for them and this too became sandy. The *same* female was observed to pick up a handful of wheat and drop it onto the water: the sand sank, the wheat floated and was scooped up and eaten. This behaviour was also then copied by the other monkeys. Did this female monkey use insight? How did this behaviour develop? We don't really know. What is required is close, careful and continual observation of the behaviour and this was not carried out. The behaviour of the other monkeys is more easily explained: they used **observational learning.**

4.6   Observational learning

This occurs when one animal observes another, the **model,** and then proceeds to copy what the model does, thus learning how to do something it was unable, or at least had not attempted, to do before. This does not mean that the learner completes the task successfully, practice is usually necessary. Chimpanzees have been observed to use stones to crack nuts. Young chimps copy the behaviour but it takes time to master the art.

One of the best-known examples of observational learning is that of blue tits which take the cream from the top of a bottle of milk. The behaviour was first seen in southern England in the 1920s. This behaviour quickly spread to other parts of the United Kingdom as blue tits copied others that had developed the skill to break open and peel back the foil cap. Peeling back bark is a common piece of tit behaviour which they use when searching for insects. Perhaps a break or tear in the cap was initially opened up by an enterprising blue tit which then found that it could puncture a cap itself with its beak. Other birds then copied the behaviour.

A series of laboratory tests carried out by E. Curio showed a link between classical conditioning and observational learning. He tested birds such as blackbirds. A pair of birds were tested at the same time, each one kept in a cage on either side of an enclosure (figure 4.6). The enclosure had wire mesh where it joined the cages so the birds could see inside the enclosure. In the enclosure was a rotating box which had four sections and into two opposite sections were placed two stuffed birds: an owl, which is a predator, and is mobbed by other birds, and a honeycreeper, which is a nectar feeder. The box was then rotated so that one of the blackbirds saw the owl and the other saw the honeycreeper. As soon as the owl came into view that blackbird gave a mobbing call. The other blackbird, which saw the honeycreeper, started to give the mobbing call in response to the alarm call of the other bird. In subsequent tests, the second bird gave the mobbing call whenever a honeycreeper appeared. Curio later showed that a blackbird could learn, by a similar training progamme, to give the mobbing call to any innocuous object, even a detergent bottle!

Figure 4.6 Arrangement of animals and apparatus used by Curio in his experiments.

At the start of this chapter, we drew attention to a blue tit which had learned to haul up a peanut fastened to a piece of string. Which form of learning do you think the blue tit used to solve this problem? We can rule out habituation and classical conditioning. It is possible that operant conditioning was involved, certainly the reward of high energy food on a cold winter day is significant. If the bird had any experience of string, or other titbits suspended like this, then latent learning could be used. Perhaps the most likely forms are insight and observational learning, with trial and error refining the task. If observational learning was important, we would still need to determine how the first blue tit (or other bird) mastered the trick of course!

4.7 Memory

Memory refers to the process by which stimuli, events and experiences are recorded and stored in the brain of an organism and which can be retrieved, when needed, at a later time. Memory allows learning to take place.

There are three stages in the memory process:

- **acquisition or encoding** – the stimulus, event or experience is changed into a form which can be received and recorded by the organism's memory system;
- **storage** – the material, after processing, is held and is available for subsequent use when needed;
- **retrieval** – the material stored is made available to the organism.

If memory fails it may be due to a fault in any one of the stages. Scientists have been investigating human memory for over 100 years. It is only more recently that they have investigated the characteristics, capacity and performance of animal memory.

A study of two species of ants was reported in 1994 in which individuals were reared in single-species or mixed-species groups. After three months, the mixed-species groups were separated and reared in same-species groups. After periods of separation, lasting from a week to six months, the ants were tested to see if they could recognise their former nest mates. It was found that they could and they were never aggressive towards them. It seems that ants can learn and memorise chemical cues that allow them to identify nest mates. The adaptive value of this is that they do not behave aggressively to, or even kill, nest mates who are usually their close relatives.

A number of species of birds, for example the marsh tit (*Parus palustris*), store seeds. They store the seeds in a variety of different places over quite a wide area and so a good memory for their store sites is beneficial. In one study, carried out in an aviary, the researchers provided 97 holes drilled into tree branches which the birds could use as store sites. The birds were tested individually. After a bird had stored 12 seeds it was removed, and two hours later it was allowed back into the aviary. The scientists found that the birds were very accurate in finding the seeds they had stored. In the second part of the experiment the researchers moved the seeds the birds had stored and put them in different holes. When the birds were allowed back they searched for the seeds in the holes they had initially used, showing that they were not using cues from the seeds, such as smell, to find the food. In follow-up experiments the same researchers gave marsh tits the opportunity to store more seeds after a similar two-hour delay. If the birds recalled where they had previously stored seeds they ought to go to different holes to store the additional seeds: they did!

It may be that food-storing birds, like marsh tits, have memory systems and brain mechanisms that have evolved to be good at making associations between a store site and its surroundings. They may be able to encode information in greater detail, or they may have better neural circuitry, than other birds. (If so, this may be another example of the interaction between the genetic and learned components of behaviour which we noted in chapter 2.) Birds spend quite a lot of time at store sites ensuring that they pack the seed in tightly and so this time may allow them to store information about the site at the same time that they are storing the seed. These birds also seem to be able to keep their spatial memory systems updated and this may mean encoding information for a large number of store sites. Simply because this type of highly specialised spatial memory might not be particularly useful for humans does not mean that we should underestimate the amazing memory skills of these birds.

Similar studies have also been undertaken with mammals. Kangaroo rats (*Dipodomys merriami*) bury seeds. However, since the stores of seeds (also known as *caches*) are undefended, other rats can find them and eat the seeds. In one study, rats were allowed to bury seeds and then had access to the area 24 hours later. During the one-day interval the researcher removed the odour from all the cache sites by thoroughly cleaning the whole area. She found that the rats could remember where the cache of seeds had been buried and could find the sites even when the odour was removed. In a second set of experiments she noted how well a cache owner did compared with a rat that had not previously had any access to the area (a naïve rat). She found that the naïve rats found far fewer sites than cache owners. The spatial memory of a kangaroo rat trying to locate its known cache again thus gives it an advantage over a rat that just searches for buried seeds.

Animals that have the capacity to remember the location of food hoards clearly benefit since they reduce the time they spend searching for food. Such behaviour confers another advantage on them too; it reduces the time they are exposed to predators.

4.8   Intelligence

In spite of the fact that we all think we know what intelligence is, it has proved elusive to define. Its study has also created controversy. For decades, there has been heated debate concerning the impact of heredity and environment on human intelligence, while the results of human tests have been used to support political viewpoints.

How can we assess the intelligence of animals? Research has shown that there are a wide variety of types of intelligence and we must also take into account the ecological and environmental influences on the animal when trying to gauge intelligence. Usually we try to assess intelligence by noting an animal's ability to learn or solve a problem, or both. Not surprisingly, animals that are higher up the phylogenetic scale are generally credited with greater intelligence.

The usual methods of assessment involve studies of the brain or the behaviour of the animals. Brain studies of animals reveal that, like humans, animal brains have areas linked to specific functions, such as the production of song. Primates have one hemisphere of the brain that is dominant for certain activities.

The behaviour of animals reveals that they do show intelligent behaviour. Use of tools has often been assumed to indicate intelligence though, when more careful observations are carried out, these acts are frequently found to show processes such as operant conditioning rather than intelligence. Thus it is quite unusual for a tool-using animal to use more than one tool, or to use it in a number of different ways. However, many animals have been observed to use tools, some showing considerable

Figure 4.7 Tool use in: (*a*) chimpanzee, (*b*) Egyptian vulture, (*c*) crow, (*d*) song thrush, (*e*) lammergeier, (*f*) elephant, (*g*) woodpecker finch.

dexterity with their chosen implement (figure 4.7). Some of the well-documented cases include:

- chimpanzee – uses a stone to crack open a nut;
- Egyptian vulture – uses a stone to break open a large egg;
- crow – drops whelks from a height to break open a whelk shell;
- song thrush – hammers snail shell on a stone to expose the snail's soft body tissue;
- lammergeier (or bearded vulture) – drops a bone from a height to obtain marrow from the bone;
- elephant – uses a twig or branch to scratch itself, remove parasites and even to hurl it at humans;
- woodpecker finch – uses a cactus spine as a probe to search for caterpillars in bark or in tree crevices.

Studies of tool use in zoo animals, especially chimpanzees and capuchin monkeys, have shown that they may possess some general skills in using objects in their environment which may be capable of transfer from one situation to another. This suggests that they are demonstrating intelligence.

So intelligent behaviour is not only the preserve of humans. In situations that are 'normal' for the organism an animal can demonstrate intelligent behaviour, behaviour that requires some thinking and planning, even formal reasoning in some cases. When we try to assess animal intelligence it is important to select criteria that are relevant to the particular animal in question: the test must use objects and an environment the animal is familiar with. If we can devise appropriate tasks we may find that animals have greater intelligence than we realise. Hens, for example, show some ability to learn by watching the behaviour of other hens on a video link!

As with humans, there is variation in the intellectual capacities of the individuals of any species. There are advantages, presumably, for individuals with greater intelligence. These advantages may be the ability to learn quickly and efficiently to find food, to use tools, to adapt to a changing environment, to outsmart predators and to employ strategies to acquire greater social status.

FIVE

Obtaining food and avoiding being eaten

On a day-to-day basis most people in Britain do not spend much time obtaining food. If we are hungry we probably go to the kitchen and search in the cupboards and refrigerator for something to eat. If the supply of food is limited we go to a nearby shop or supermarket. In a country like ours, with an assured year-round supply of a wide variety of goods, obtaining food is not a problem. Even less of a problem is the possibility of being eaten!

Obtaining food and avoiding being eaten are, however, the daily priorities for the vast majority of wild animals. Feeding is preceded by foraging. **Foraging** refers to all behaviour that is associated with getting food and for which the organism must search or hunt. **Feeding** is the behaviour that is associated with the ingestion of food, which may require constraint and manipulation.

Some animals are described as being **specialist** feeders, others as **generalist**. To some extent, all animals are specialists since most take only a small fraction of the potential food items available. However, some animals do feed on a very restricted number of items. These specialists include the koala bear which feeds exclusively on eucalyptus leaves and the giant panda which feeds almost exclusively on bamboo. Generalists take a broader range of foods; rats and foxes are examples. An advantage and disadvantage of each feeding strategy is outlined in table 5.1.

Having decided to feed, one of the key questions an animal faces is **where to search**? How an individual answers this question is very

Table 5.1 An advantage and disadvantage of the generalist and specialist feeding strategies.

	Generalist	*Specialist*
Advantage	Can switch from one food to another as conditions dictate	Is well adapted to the food item in terms of finding it, handling it and digesting it
Disadvantage	The skills of finding and handling the food item are not well developed	What does it eat if the food item declines in abundance?

important. In general, the individuals that solve this problem most efficiently will do best, that is have the highest inclusive fitness (see chapter 1). This is because they spend less time and energy finding their food, in relation to the amount of food they require, so that more time is available for caring for young, hiding from predators, establishing a territory, finding a mate, etc.

Foraging, like any other piece of behaviour, can be understood in terms of the benefits and costs involved. This is a key concept in behavioural ecology and is the core of **optimality theory**. This concept is based on determining the costs and benefits of any action and suggests that the behaviour we observe will have evolved over time to be **optimal**, that is the best under a given set of conditions. The optimal solution is to behave so that the benefits outweigh the costs by the greatest amount. Natural selection determines which of a number of behavioural alternatives is best.

The concept of optimality has been widely applied in the area of foraging behaviour and many studies have been carried out to determine the optimal foraging behaviour for an organism under a particular set of conditions. These conditions might include: which food to look for, how the nutritive value of one food item compares with others, if the food is abundant or rare, if the food item is easy to catch and subdue and if the food is to be shared with others.

5.1 Optimal foraging

Whatever the food item, it is seldom abundant and everywhere: it is usually distributed in **patches**. Patches are not distributed in a regular manner and they vary in quality, so foraging efficiently is not easy. It is made more difficult by the fact that each animal is competing with others, of its own or of a different species, and it must also be aware of predators as it forages.

In order to select a profitable patch the animal must sample a few in order to find a 'good' one, or ideally the 'best' or 'optimal' patch. How many patches should it sample? How long should it stay in a patch? These are questions of real importance and it is vital for animals to make the 'right' decision, whether or not that decision is made consciously.

How many patches should an animal sample? A number of studies of foraging have been carried out in the laboratory using birds. Typically, the researchers have given the birds the opportunity to forage in a number of areas in an aviary. The density of food in each area differs so that patches are created. The observers then record the amount of time spent in each patch. In a study of great tits it was found that they spent most time feeding in the section where the food density was highest but they did occasionally sample the other areas. By doing this it might seem that the birds were choosing a sub-optimal solution. However, when the experimenters reduced the density in the 'best' patch, the birds quickly switched to the next best. Thus it pays to visit other patches in order to gain information about the availability of food there in order to utilise this source if the food density in the 'optimal' patch falls.

When should an animal leave a patch? If an animal feeds for any length of time then the number of food items is reduced and the quality of the patch declines. At some point the animal would do better by leaving the patch and moving to another. How does it decide when to move? To make this decision the animal needs to know what is the average amount of food available in all the patches in its area, how much is available in the patch it is currently in, and how long it will take and how much energy it will use in moving to a new patch. A graphical solution, or model, can be used to predict when the animal should move: the model is called the **marginal value theorem** and is illustrated in figure 5.1.

The animal's rate of gain in energy from eating is initially high, and highest at the point of origin, as it encounters a high density of food items. As it eats the food, the density decreases and the animal's intake of food decreases too. It should therefore leave the patch at time t since, after this, the rate of intake of food energy is less than the average value for the environment, that is the 'marginal value'. (This ignores the cost of moving to another patch. In fact the cost of commuting to the next patch can be important. This is particularly so if the cost of getting to the next patch is high, perhaps because the patch is distant.)

Is there any evidence that animals behave as the model predicts? The answer is yes, but they don't behave exactly as predicted. Factors that the model does not take into account are important too. Animals have other priorities. For example, mate guarding was found to influence foraging decisions in wild male baboons (*Papio cynocephalus*). Researchers studying baboons in southern Kenya found that males travel shorter distances to forage when mate guarding and also had shorter feeding bouts. When mate guarding, it seems that males are prepared to pay the costs of a reduction in

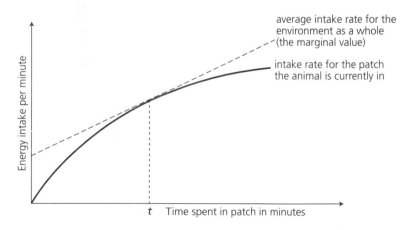

Figure 5.1 Predicting when to leave a patch: the marginal value theorem.

energy intake since the benefits of guarding, that is copulating with females and producing offspring, are significant.

A neat foraging environment was created in the laboratory by two researchers in 1992 when investigating foraging decisions in a species of wolf spider, *Schizocosa ocreata* (figure 5.2). They were trying to determine the way in which spiders assess the quality of a patch and which type of stimuli are influential. Sixty female spiders were placed, one by one, in the centre of the apparatus and allowed to move freely between the four different patches for one hour. The stimuli were 12 immature crickets in each patch separated from the spider by an acetate sheet. The researchers measured how long a spider spent in each patch by videotaping its behaviour from above. Would the sensory cues from the crickets determine how long a spider spends in each patch?

The patches varied in the type of sensory information the prey provided: in one patch there was no prey (control), in another visual stimuli only, in another auditory/tactile (that is, *vibrational*) stimuli only and in the fourth both visual and auditory/tactile stimuli. They found that spiders stayed longer in the patch with visual and auditory/tactile stimuli and the patch with visual stimuli only (figure 5.3) than either of the other two patches.

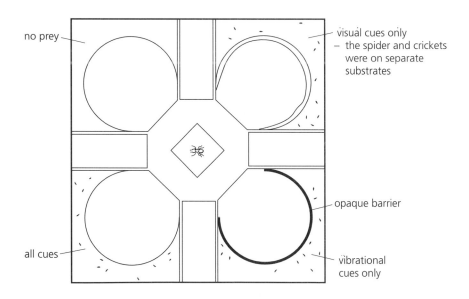

Figure 5.2 The artificial environment used in the experiments. Apart from the top left quadrant, 12 immature crickets were placed in each chamber. The sensory treatments were as follows, clockwise from the top left: (1) control chamber, no crickets; (2) visual stimuli only; (3) auditory/tactile (i.e. vibrational) stimuli only (an opaque barrier was placed between the spider and the crickets); (4) visual and auditory/tactile together.

Optimal foraging

It is not surprising to find that this species of wolf spider, which hunts its prey in leaf litter, makes use of visual stimuli in particular. This species uses a 'sit and wait' strategy – when a suitable prey item comes along the spider makes a quick dash to grab, subdue and eat it. (This species of spider has a visual range of 30–40 cm.)

Spiders must have a decision rule to decide when to leave a patch and go and hunt elsewhere. This decision rule could be based on time. If after a certain time a suitable visual stimulus (that is, a potential prey item) does not come along the spider moves on to another patch.

The assumption that underpins optimal foraging models is that animals behave optimally, that is the animals forage in the most efficient manner possible, not just merely adequately. This is because the optimal foragers benefit considerably from spending less time and less energy searching for food. As a consequence optimal foragers can devote more time to establishing and defending a territory, building a nest, guarding a mate, and so on, and so will leave more offspring than those animals which do not forage optimally. Of course, this does not mean that all individuals in a population are equally adept at finding food. Some will do better than others. However, over countless generations the forces of natural selection have led to individuals that forage most efficiently being the ones that survive and produce most offspring.

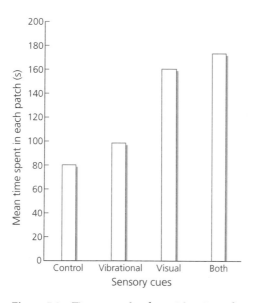

Figure 5.3 Time spent by the spiders in each patch (s). The treatments were: control (no crickets); auditory/tactile (vibrational) stimuli only; visual stimuli only; both visual and auditory/tactile stimuli from the crickets.

5.2   Factors influencing foraging decisions

Apart from which patch to feed in and when to leave it, there are many other factors that influence individuals as they look for food. Some of these factors are outlined below.

Prey size

If an animal has a choice of two prey items, one large and one small, does it always go for the larger? The larger prey item provides the greater net gain of energy; however, since prey are invariably reluctant to be caught and eaten, the predator must spend some time in subduing and handling the prey before it eats it. Small prey therefore may sometimes be more profitable. In a laboratory study of shore crabs (*Carcinus maenas*), two researchers offered them mussels (*Mytilus edulis*) of different sizes to eat. They found that the crabs didn't always take the largest, in fact they took a range of sizes. Small mussels are sometimes taken if the time to search for food is great, that is if prey abundance is low.

Constraints on diet

A study of moose (*Alces alces*) showed that physiological factors are important in determining the daily diet of these animals. Three constraints operating on moose are the size of the rumen, the energy content of the food and its sodium content. Moose feed in both aquatic habitats (where the plants have a high sodium content) and woodland habitats (where the plants are low in sodium but high in energy). The researchers took measures of rumen size and the daily needs of energy and sodium and were able to predict what the trade-off between energy and sodium should be. They found that the diet of moose was such as to ensure that their daily requirement for sodium was met, they can't just maximise their energy intake.

Dominance

In group-living animals there is usually a **dominance hierarchy,** a ranking of individuals according to their social status. One study of feeding competition between red deer hinds (*Cervus elaphus*) was carried out on the Isle of Rhum, Scotland. The scientists focused on aggressive interactions, approaches and withdrawals between females and found that the subordinate females were much more likely to stop feeding and move away if a dominant hind approached. They also found that subordinates took fewer bites of grass as the distance between them and a dominant female decreased.

Learning from other animals

Learning from others is possible and one laboratory study showed how parental choice is influential. This was a study of house finches (*Carpodacus mexicanus*) reported in 1996. Adult pairs of nesting finches were offered equal amounts of canary seed and oats during the experiments. A chemical, methiocarb, was added to the oats which were fed to some pairs. This made the birds sick so that they avoided eating oats. This treatment continued until the eggs hatched. At this point all the adults were offered untreated oats and canary seed, which they fed to the young by regurgitation. The researcher found that the offspring of pairs who had previously been given oats treated with methiocarb were fed very few oats compared with adult pairs whose food had remained untreated. For the next 5–6 weeks the young birds were fed by their parents but after fledging, when about 6 weeks old, they were separated from the adults and raised in other cages. Untreated oats and canary seed were available to them in equal quantities each day and the mass of seed and oats eaten by the birds was recorded.

It was found that the birds which had been raised by adults that had avoided oats ate less oats than birds raised by adults that had eaten oats. The adult birds had learned to avoid oats because of their prior nauseous experience and their offspring had learned to avoid oats too. How did the young birds learn to avoid oats? They could have learned what to eat by watching their parents and noting which containers they ate from in the cage. They might also have followed their parents to the feeding area, when old enough to be able to do so, and noticed that their parents largely ate canary seed. Thus the food avoidance learning of the parent finches had a direct effect on the diet of their offspring.

So, a consideration of the costs and benefits of foraging allows us to determine what would be the optimal solution for the animal. How it forages optimally depends upon factors such as prey size, constraints on diet, dominance and learning.

5.3   Strategies used by animals searching and hunting for food

The food of herbivores does not take direct avoidance action to prevent itself being eaten. This does not mean that most plants encourage animals to eat them. A number of plants, such as milkweed and ragwort, build up poisons which inhibit animals from eating them. However, predators often have to use specific strategies to get their food.

Cheetahs painstakingly stalk and then use a blistering burst of speed to catch their prey, such as Thomson's gazelles (*Gazella thomsoni*). Once it bursts into action, a cheetah can sprint at 95 kilometres per hour for 30 seconds or so. During this period, the cheetah tries to trip the gazelle as it runs away. Gazelles, of course, do not acquiesce but sprint away too. They swerve as they run to make it more difficult for the chasing cheetah. Gazelles have greater stamina and so if the cat does not catch them in the first few hundred metres it gives up. In fact it is so exhausted that it often takes up to an hour for it to recover and be ready to hunt again.

A 'wait and pounce' strategy is used by some predators, for example moray eels. For most of its time a moray eel waits in a crevice or beneath a rock on the sea bed. When a fish comes close the eel darts out, grabs it, retreats to its hideaway and consumes the meal. This strategy is not confined to aquatic predators. Some spiders build an underground burrow which has a trapdoor. The spider waits just inside the door and then, as prey walks by, it opens the door quickly and pounces on its hapless victim. It then drags it into the burrow to eat.

Lures are used by some predators to capture their prey. The bolas spider (*Mastophora* sp.) uses a sticky blob of silk on the end of a thread as a lure. The spider swings this blob around. The swinging blob draws moths, particularly as the silk droplets contain chemicals which are like those that female moths emit to attract males. When the moth alights on the swinging sticky blob it is hauled up by the spider and eaten. This species is also a good example of a specialist feeder since the spider feeds on male moths of just one or two species.

Some group-living predators, such as African hunting dogs (*Lycaon pictus*), combine their individual hunting skills and co-operate together. They are more effective if they do so. Often the prey are faster than an individual dog but by functioning as a unit they ultimately wear down the prey, since the dogs have greater stamina. Also, by hunting as a group the dogs can kill larger animals than they could manage as an individual. Of course, this is offset by the fact that the food is shared, but the benefits of sharing outweigh the costs. This does not mean, however, that all individuals in a group of social animals get an identical share of the food, higher ranking animals usually do best.

One example of co-operation involves two different species. The African honeyguide (*Indicator indicator*) is a bird that feeds on insect larvae and the waxy comb of honey bee nests. Since the nests of African honey bees are in crevices in trees they are mostly inaccessible to birds. The solution is for the bird to lead another animal, the honey badger (*Mellivora capensis*), to the nest site. The bird does this by making a noisy, distracting call and flying close to the badger. If the badger follows, the bird flies on again, continuing the call, and eventually guides the badger to the nest. The honey badger has sharp claws and powerful limbs and so can open up the nest. After it has eaten, the badger moves away and the honeyguide can share the spoils. Honeyguides also lead humans to bee nests for the same purpose.

5.4   Selecting what to eat

Selecting what to eat can depend on having a **search image** for the particular prey. A search image is used when an animal actively seeks one particular item of food whilst ignoring others. A search image for a prey item is extremely valuable for a bird that eats camouflaged prey, for example blue tits searching for moth caterpillars. If the blue tit develops a search image for the item then it becomes quicker at spotting others. The same principle operates with humans too. If you are searching for moth caterpillars in bushes or trees (the sort of activity that one of the authors of this book frequently indulges in!) and you suddenly spot one, you can then often spot others; we say that you have 'got your eye in'.

Predators are usually selective when making a decision about which animal to attack if the prey are in a group. They don't make a random choice – it is usually the older or younger or weaker or less vigilant animals that are selected. What are the advantages and disadvantages of selecting these animals as the focus of the attack (table 5.2)?

Table 5.2   The advantages and disadvantages of taking weaker prey.

Advantages	*Disadvantages*
They don't have the stamina of fitter individuals.	Older animals may not supply such nutritious food.
The predator may be able to get closer to them before launching an attack.	They may be diseased.
They don't run as fast and so are easier to catch.	They may be smaller so there is less to eat.
They don't fight back very effectively.	

A study of cheetahs stalking Thomson's gazelles in the Serengeti National Park illustrates how cheetahs select less vigilant animals. The researcher noted occasions when a cheetah approached two gazelles of the same sex and which were within five metres of the other. As it stalked them the cheetah observed the two animals carefully. The researcher recorded the time that each gazelle spent feeding and scanning during the cheetah's stalk. She found that the less vigilant member of the pair was chased on nearly 90% of the occasions. The cheetah increases its chance of catching the less vigilant gazelle since it can be approached more closely before it runs.

Some animals hoard food in times of plenty and draw on this supply, or **cache**, when food is scarce. For this to be effective the animal must have sufficient memory storage and retrieval so that it performs better than individuals of other species that don't use food stores. We have already noted a number of examples (see pages 55–56). Work with black-capped chickadees (*Parus atricapillus*), American members of the tit family of birds, suggests that in these birds the cache sites tend to cluster in a particular direction, thus making it easier to find the seed and also reducing the animal's requirements for a high memory storage and retrieval system. In the wild the black-capped chickadee cache sites are pilfered by animals of other species. Do the birds avoid the use of pilfered sites again? It seems not! Previously pilfered sites are not avoided when the birds cache subsequently. Presumably the bird's memory system does not allow information about the success of every site to be stored. Perhaps the costs of remembering 'safe' and 'unsafe' cache sites are simply too costly or perhaps this is beyond the birds' capabilities.

The red-backed shrike (*Lanius collurio*) uses a rather more unusual type of food store. Shrikes are aggressive, carnivorous birds and are also known as butcherbirds. This term explains their method of storage. Their prey are chiefly insects but include mice, lizards, frogs and smaller birds. Having grabbed and killed their prey, the birds then impale their victims on thorns. The birds can then use this store of food on a later occasion.

Another stratagem is to 'grow your own food'. There are only a few examples of animal agriculturists and probably the best known are the various species of leaf-cutting ants. These ants grow fungus gardens inside their nest chamber. Forager ants go out in search of leaves which they cut into small pieces to carry back to the nest. Inside the chamber the pieces are chewed. This encourages the growth of fungi, which is then used to feed adults and larvae.

Occasionally, some animals have been observed to take 'food supplements'. This behaviour has been noticed in some species of macaw. The birds gather in groups on soil exposures, for example, the banks of rivers, and gnaw at the soil. It is thought that the minerals in the soil are effective in neutralising toxins which the birds consume as a result of eating seeds.

5.5   Avoiding attack by predators

Remaining vigilant is one of the most obvious ways of avoiding attack. Vigilance is enhanced if the animals group together, since there are many eyes, ears and noses that can detect the approach of a predator. In the autumn, winter and early spring, flocks of black-headed gulls are frequently seen on school sports fields. By foraging in a large flock the birds get an earlier warning of approaching danger, perhaps a dog, than if they forage on their own.

A very specialised way to enhance vigilance in a group of animals is to have one individual, the **sentinel**, who remains alert and warns others in the group if danger threatens. Usually adults in the group take it in turns to act as the sentinel. Such a system has been described in meerkats (*Suricata suricatta*), see figure 5.4. These are mongoose-like animals of southern Africa which have a very highly developed social system. Whilst most members of the group are foraging, one or two sentinels are located in prominent positions, for example on the top of a bush or a termite mound, to keep an eye open for predators. If a predator is seen, an alarm call is given and the group heads for cover or their underground burrow.

A common way to reduce the likelihood of attack is to rely on camouflage. Many predators rely on their eyesight when hunting and so if prey are coloured to match their background they are less easily detected. Of course, disguise is usually aided by keeping still. Natural selection favours prey that are motionless and well camouflaged through **cryptic colouration** (that is they match their background very closely). However, prey animals eventually need to move to search for food, for a mate or to protect their young. There are many examples in the natural world of animals looking and behaving like objects in their environment, such as leaves, twigs, bark and even bird droppings! The prey must ensure that they select the substrate for which they are well camouflaged. Moths are mostly nocturnal and spend the day at rest on walls, tree trunks, fence posts, leaf litter, grass or other types of vegetation. They select their resting site very carefully so that they blend in with the background. Whilst camouflage is a common means of deception, prey animals often have a back-up stratagem in case cryptic colouration fails. Stick insects, for example, drop to the floor if they are attacked.

Some animals use colour in a different way: they aim to be highly conspicuous. These animals send out colour signals indicating, through vivid and often patterned colours, that they are poisonous (arrow frogs and coral snakes), taste foul (Monarch butterfly) or sting (wasps). As a consequence, predators soon learn to avoid them. Some other species take advantage of this and are similarly coloured, even though they are harmless: king snakes look rather like coral snakes and clearwing moths look like wasps. These animals derive an advantage by mimicking a potentially dangerous animal. This is known as **Batesian mimicry**. In **Müllerian mimicry**, several species

Figure 5.4 Meerkats in their characteristic vigilant posture, sitting back on their haunches.

look alike yet each is unpalatable (figure 5.5). Can you see an advantage to each species of this? (The advantage to each potential prey species is that predators only need to learn to avoid one colour pattern, rather than many.)

Certain animals surprise a predator by their behaviour. For example the puffer fish inflates its body to become larger; the eyed hawk-moth reveals two large eye spots on its hindwings if disturbed so that it startles the bird; the larva of a lobster moth vibrates its long legs so that it resembles a spider; an opossum will pretend to be dead if attacked; some species of skink can lose their tail if attacked (the tail is regenerated in a few weeks).

(a)

(b)

Figure 5.5 (a) An example of Batesian mimicry. The clearwing moth (left) looks very like the wasp (right) but has no sting. (b) An example of Müllerian mimicry. *Heliconius isabella* and *Mechanitis isthmia* are found in the same areas. They have similar markings and are both unpalatable.

5.6   Avoiding capture by predators

Most potential prey animals, ungulates for example, cannot afford to spend time motionless but need to move about searching for food. In the Serengeti in East Africa there may be over one million wildebeest. These are large black-brown herbivores that do not rely on camouflage for survival. They combine alertness and running speed to escape from predators, such as lions, leopards, spotted hyenas and hunting dogs. Prey species like wildebeest, gazelles, impalas and zebra rarely keep their heads down for long but look up to scan for danger. They also look to see what other herbivores are doing. Keeping vigilant thus reduces the risk of attack.

Another important strategy that is used by prey animals is to coexist with others in a group. By foraging in a group the animals increase the chance that one of them will spot a predator and take the appropriate action, thereby warning the other group members.

Foraging in a group also means that the animals can work together to nullify the threat from predators. If the group is sufficiently large, then predators may not attack. This effect has been recorded in musk oxen. If a group is attacked by wolves, the musk oxen circle closely together. The adults face outwards with the young in the centre of the circle. If the adults stay in a tight circle, they are often effective in repelling the attack.

A study of goshawks (*Accipiter gentilis*) and wood pigeons (*Columba palumbus*) illustrates how group size affects the survival chance of an individual pigeon in a flock when a goshawk attacks. The research showed that if a wood pigeon was attacked when it was on its own, the chance of the goshawk being successful was about 80%. However, if the pigeons were in a flock the chance of success decreased rapidly, being less than 10% if the flock numbered over 50 (figures 5.6 and 5.7).

A study reported in 1995 concerned laboratory work in which cichlid fish (*Aequidens pulcher*) were the predators and guppies (*Poecilia reticulata*) the prey. The study showed that the cichlids were actually more likely to attack larger shoals. However, although the risk of attack to the guppies was greater in a larger shoal, their risk of being taken was less than if they were in a small shoal. This may be because, when a shoal is attacked, their sudden flight and scattering may confuse the predator and this allows the prey to escape.

Of course it might be possible for an individual animal in the group to cheat and carry on feeding without scanning. This does not seem to occur, though we don't know exactly how cheating is prevented, or at least minimised.

Another advantage to being in a large group is that *any individual* is proportionately less likely to be attacked by the predator. This decrease in the chance of an individual being attacked as group size increases is termed **dilution.** However, the chance of each individual in the group being

Figure 5.6 A goshawk and a wood pigeon.

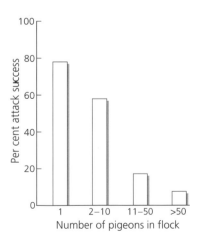

Figure 5.7 Percentage success rate when goshawks attacked flocks of wood pigeons.

attacked is not identical – those in the group nearest the predator's attack have a greater risk, as do weaker, older, younger, diseased and injured members. Predators thus tend to focus on certain individuals in the group in order to reduce the effect of the confusion. If selection of likely victims does occur then we might expect prey animals to indicate to predators that they are fit and healthy. Some species of antelope appear to do this. When the predator has been seen, the antelopes sometimes leap high into the air. This action is termed **stotting**, and it may demonstrate to the predator that they are in prime condition and will escape, or out-run them, if attacked. A predator may then select another target to attack, perhaps a less healthy or less vigilant animal.

Some animals take more direct action to repel a predatory attack; one example is the bombardier beetle. If attacked, the beetle produces a chemical spray, rather like a chemical explosion. This is expelled from the body with considerable force and at boiling point (100 °C) which is very effective at inhibiting further attack. Skunks are also well known for producing a smelly noxious fluid that is squirted at the attacker.

Another form of direct action is the deception employed by a plover to lure a predator away from its nest. If a nesting plover sees a predator, such as a fox, it moves quickly off its nest until it is some distance away, whereupon the bird feigns injury by mimicking that it has a broken wing. The wing is held in an unusual manner, being dragged along the ground. The bird moves further away from the nest and may even give a convincing show of trying to fly while injured. The fox is distracted by this display and moves towards the 'helpless' victim. Needless to say, if the fox gets too close the plover flies off and eventually returns to its nest when the predator has left to seek other, less distracting, prey.

SIX

Courtship and mating behaviour

In this chapter we shall examine how natural selection has shaped the courtship and mating behaviour of animals, why sex has evolved, how males compete for females and how females choose males. We will also look at how humans choose their mates. We begin by looking at courtship and mating behaviour in a species of gull.

6.1   Life in a gull colony – an example of courtship

Lesser black-backed gulls (*Larus fuscus*) are large (about 55 cm in length) black and white birds with a yellow bill and yellow legs. They nest in colonies, frequently in grassy areas or sand dunes. The male bird is the first to arrive at the breeding ground and seeks to establish a nesting territory in a suitable area. In fact, many male lesser black-backed gulls claim the same territory year after year. The male competes with other males for space and, once he has established a territory, he defends it vigorously. To attract a mate, a male calls from the nesting area. Females flying overhead may fly down to inspect the male and his territory. If she is impressed by him, and the nesting area is suitable, she may stay. Males are in a highly aggressive state, however, and so, to ensure he does not attack, the female approaches him submissively.

Over the next few days, the male feeds the female on the nesting territory, though each bird also leaves it to find food for itself. The time during which the male feeds the female is very important as it serves to strengthen the bonds between the birds and decreases the aggressiveness of the male towards her. To get food from the male the female approaches submissively with her head held low and gives a begging call, rather like the call of a chick. The male then regurgitates food to her, which builds up her body reserves in preparation for egg production and incubation.

Copulation, which occurs quite frequently, only happens after a number of days of feeding. During the pre-copulatory period the female assesses how good the male is as a provider of food and as a defender of the territory. Both these qualities are vital if young are to be reared successfully. The investment in the future is considerable for both birds: if four eggs are laid and hatched, then it needs two committed parents to ensure the chicks survive to fledging and independence. Making the right choice of a mate during courtship is vital for both birds.

It is not surprising that individual male and female lesser black-backed gulls often keep the same mate for many years. Keeping the same mate reduces the length of courtship since the birds are already familiar with each other and each bird knows how successful their partner is as a parent. Usually the pair improve their breeding performance over the years too, that is they raise more chicks to fledging. So, staying faithful to a mate pays off. In fact, the 'divorce' rate in seabirds is quite low, one recent study of great skuas (*Catharacta skua*) found it to be only about 6% each year.

Courtship, then, is seeking out and mating with other animals. **Courtship** can be more formally defined as the set of behaviours that are associated with the sexual union of two individuals which results in the production of offspring. A successful mating means that each animal passes on copies of its genes to the next generation.

6.2   Must there always be a father?

If you look closely at most rose bushes in the spring or summer you will probably notice tiny green insects on them, especially around the tender buds. These insects are aphids (greenfly) and they feed on plant sap. Aphid females are able to **reproduce parthenogenetically**, which means that their eggs develop without being fertilised by a male. Reproduction is **asexual**. Each daughter aphid produced like this has only one parent and is genetically identical to its mother. For a female aphid, the advantages (benefits) of not needing a mate are as follows:

- there is no need to go and search for a mate – this saves time and lowers the risk of predation;
- the offspring will be identical to you – so if you are successful then your offspring are likely to be successful too, provided the habitat does not change;
- you avoid the possibility that the male you would otherwise mate with may be infertile;
- there is no need to produce males and so every aphid in each generation makes a direct contribution to population increase – this is a great advantage over just a few generations and is why aphids reproduce parthenogenetically in early spring and summer.

The main disadvantage (cost) with asexual reproduction is that the offspring do not vary genetically, they are identical to their mother. This is a problem if the habitat changes suddenly. It is for this reason that sexual reproduction in aphids does take place from time to time to provide genetic variation.

The vast majority of animals use **sexual reproduction** and so two parents are required. Presumably this is because the benefits of reproducing sexually outweigh the costs. The costs are as follows.

- Each parent is only responsible for half of the genetic contribution to each offspring and so its contribution is diluted.
- A considerable amount of time and effort may be needed to find and select a mate.
- Animals make themselves conspicuous when they seek a mate and may be caught by predators.
- There is no guarantee that the mating will result in offspring (of course, not every egg produced by asexual reproduction is guaranteed to be viable).
- In some cases, courtship can result in the death of a prospective partner – this may even occur *before* mating. In certain spiders the females sometimes grab and eat their suitors! To reduce the chances of being eaten, a male wolf spider approaches a female slowly and waves his palps in a special way to indicate his intent (figure 6.1). Death of a partner can also occur *after* successful courtship. In the case of mantids, such as the praying mantis (*Mantis religiosa*), the female sometimes eats the male as they are mating! However, the male does continue to pump sperm into the female as the upper part of his body is eaten and so mantids still reproduce sexually, albeit some males only mate once!
- Some individuals do not mate – this is often the fate of males since they need to compete with other males for mating opportunities: the victors usually mate, the losers are often unlikely to do so. In most species females rarely go unmated, if they are healthy.

Figure 6.1 Male wolf spider (*Lycosa amentata*) giving courtship signals to a female before approaching her. This drawing shows two elements in the sequence of visual signals that a male sends to a female spider.

- For animal species which show biparental care (that is two parents rear the offspring) there is the danger for females of being deserted by the male after mating has occurred. For males there is the danger of cuckoldry, that is rearing the offspring of another male who has also mated with the female.

The benefit associated with sexual reproduction is:

- evolution is faster since sexual reproduction involves combining genetic material from two individuals, thus favourable combinations can evolve more rapidly than is possible in asexual reproduction.

Given that most animal species have two sexes, what is the basic difference between the sexes?

The essential difference between females and males is that females produce a few, relatively large sex cells or gametes (eggs), whilst males produce massive numbers of tiny sex cells (sperm). It follows that it is theoretically possible for one male to fertilise a very large number of females. For such a male there would be a huge pay-off to his promiscuity since he would leave a large number of offspring. Other males are unlikely to acquiesce, however. This means that males generally compete with each other to mate with the females. Females, on the other hand, need to take great care of their limited number of eggs and so they put considerable effort into choosing the 'best' male to mate with. Crucially, what follows from this is that females are the scarce resource and as a consequence they may be able to select which male to mate with.

Males consequently compete for access to the females. Competition between females does occur but is much less severe. There is also conflict between the sexes over courtship, mating and parenting.

6.3   How do males compete?

Males compete in two ways: either by competing with each other in trials of strength and stamina, known as **intrasexual selection** (selection within the male sex); or by appealing to females using behavioural and ornamental displays, known as **intersexual selection** (selection of males by females). Let us consider an example of each type of competition.

Intrasexual selection

Red deer (*Cervus elaphus*) males illustrate intrasexual selection. In the autumn breeding season, female deer (hinds) assemble in groups of up to 30: such a group is called a **harem**. Individual males (stags) attempt to defend these female groups from other males, so as to mate with the adult females. Only large mature males can hold a harem. The effects of millions of years of selection have resulted in males which are almost twice the mass of females and possess large antlers, potentially lethal weapons. Male–male competition occurs in a number of ways. A male holder of a group of females may, by virtue of his physical bulk and visible antlers, dissuade other males from challenging him. However, studies have shown that males don't just rely on the visual signal of body size, they back it up by roaring several times per minute for most of the day! Why do you think males engage in such exhausting bouts of roaring?

Roaring may be sufficient to dissuade challengers. However, other males may begin roaring and challenge the harem holder. Stags can assess the physical condition of other stags by their rate of roaring and perhaps depth of roar. If a challenger believes his roaring rate is greater than, or similar to, the holder he approaches and the two stags visually assess each other from just a few metres away whilst they are both walking: this sequence is referred to as **parallel walking**. If this does not dissuade the challenger the two protagonists lock antlers and engage in a pushing contest to assess which is the stronger (figure 6.2). Each also tries to twist his opponent in an attempt to dislodge him or catch him off balance. The battles rarely last more than a few minutes and the weaker withdraws. The stronger stag then mates with the females as they come into oestrus. This example also shows that animals tend to fight in conventional, restrained ways. It is as if they have a set of rules which they each adhere to. The contests are **ritualised** and the outcome is decided through a lower level of aggression than

Figure 6.2   Two red deer stags preparing to fight.

might be the case if they had a no-holds-barred contest. Thus red deer stags could settle the outcome of the contest at any one of the stages we have outlined. Although each stag carries potentially lethal weapons in its antlers they are used as a last resort and generally in a restrained manner.

Size usually confers an advantage in male–male contests. However, this does not necessarily mean that the bigger of the two combatants always carries the day. In Weddell seals (*Leptonychotes weddelli*) males, which are smaller than females, commonly fight for access to females. They try to hold the females in harems of 2–12. The winner in male–male fights is not always the larger of the two. The smaller male has the advantage of greater mobility when fighting and this may tip the balance of the contest in his favour.

What these examples of intrasexual selection show is that in these sorts of mating systems, that is ones with strong male–male rivalry, there is little scope for female choice.

Intersexual selection

When males try to attract the attention of females using elaborate adornments, in intersexual selection, a number of males often compete together. This is nicely illustrated in the great snipe (*Gallinago media*), a bird which breeds in marshy areas in Scandinavia and Russia. Male and female snipe look almost identical except for the amount of white on the tail feathers – the males have more white than the females (figure 6.3). Males have territories on a display area or **lek**. The lek is like a parade ground over which the males strut and display their feathered finery to visiting females. A lek may

Figure 6.3 The end of a great snipe display showing how a male displays his tail feathers.

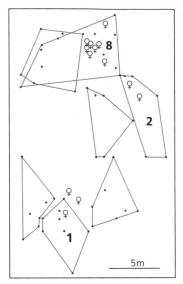

Figure 6.4 Map of the great snipe territories on a small lek (seven males) on one night. The number indicates the number of females that visited the lek on that night. The dots indicate the positions of the male on each territory and allowed the boundaries of each territory to be defined. (In this study the lek was an area of grassland just above the tree line at 500–900 m.)

have a number of territories; one studied in Sweden had seven (figure 6.4). If a male does not gain a territory he is unlikely to mate.

In one study, researchers netted some male snipe and artificially increased the amount of white on the tails of one set of birds and darkened the equivalent area on another set. They found that females preferred males with larger areas of white on their tails. It is known that the amount of white increases with age and so the females are preferentially selecting older males. Can you suggest a reason why females prefer older males?

In cases where a female will visit several males before selecting a mate, it is interesting to think about how she decides on the number of males to visit. The easiest decision-making rule would be for her to mate with the first male she sees. This might even be the safest rule, as there is probably a greater risk of predation at a lek, but studies do not seem to suggest that this occurs in natural populations. Why not? Presumably, it carries costs which exceed the benefits of ease and safety. The first male a female comes across might be undernourished or have a high parasite load or be infertile. It would be best for a female if she made a decision regarding male quality on the basis of a characteristic that indicates not only his physical state but also his genetic make-up. What should females do and is there a general rule that they follow? The answer is we don't really know, but a study of peacocks showed that peahens visit, on average, three males before making a choice. They also seem to choose on the basis of the number of eye-spots in the tail and the condition of the tail. Clearly there must be some upper limit on how many males to visit. If a female spent all her time inspecting every possible mate before choosing she would not have time to rear young in that breeding season.

Other forms of male competition

Males also compete in other, more subtle, ways than fighting. If a female mates with two males then the sperm from both will compete to fertilise her eggs; so there is **sperm competition**. Males can adopt tactics which increase the likelihood of their sperm fertilising a female's eggs. The following are some of the tactics employed by males.

Removing the sperm of competitors

If this is to be successful, we can imagine that a male must have a suction pump or scoop which he carries around for just this purpose. It may seem unlikely, but one species of dragonfly (*Calopteryx maculata*) has just such a mechanism. The penis of the male has evolved a scoop-like structure which makes removal of sperm from the female easier. After he has removed the sperm of any males that had previously mated with the female, he deposits his own sperm.

Using a copulatory plug

Male garter snakes (*Thamnophis sirtalis*) use this tactic. After mating, the male produces a sticky gum which seals up the female's genitalia. This prevents any other male from mating with her; it also seals in his own sperm.

Mating lock or tie

You may have observed this yourself as it occurs in dogs (*Canis lupus familiaris*). After ejaculation the male cannot withdraw from the female since the base of the penis swells. This clearly increases the likelihood of this male's sperm fertilising the eggs since no other male could possibly mate with her whilst they are locked together.

Mate guarding

The guarding of females by males can be either *before copulation* (as a female approaches the time of sexual receptivity which is the ideal time for the male to mate with her) or *after copulation* (when the male protects the female and prevents other males from mating with her).

Guarding behaviour before copulation has been studied in elephants (*Loxodonta africana*) in Kenya. When the female elephant is in oestrus she makes low-frequency calls which attract males. The calls indicate that she is close to sexual receptivity. In the early phase of oestrus, younger male elephants (25–35 years old) do mate with the female but they are unlikely to fertilise her. However, during the period of mid-oestrus, when she is in her

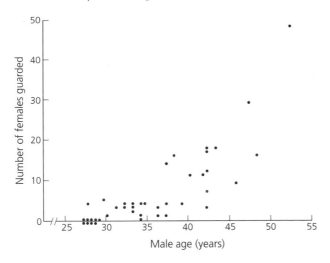

Figure 6.5 Relationship between male age and guarding success in elephants. There is a significant positive relationship between the two variables.

most receptive state, it is older males (>35 years old) who guard her and mate successfully with her (figure 6.5). To be successful in guarding, the male must be larger and more aggressive than his challengers.

Mate guarding after copulation occurs in a number of dragonfly species. The male clasps the female behind her head and flies off with her to seek suitable egg-laying sites. He stays with her as she deposits her eggs, thus preventing another male from mating with her.

Female mimicry

Most people are amazed to learn that there are insect transvestites! It might seem a bizarre tactic for a male to use but it can be a successful one.

One study found that some male scorpion flies mimic the behaviour of females. If a male is to court a female successfully he must present her with a gift of a fly for her to eat. This builds up her body reserves and allows her to lay more eggs. Some males wait in the vegetation and ambush a fly. Others try to pinch one from a spider's web – a risky venture! Some males alight on a twig and take up the body posture of a female. Another male may be convinced by this and offer the gift whereupon the mimic grabs the offering and then seeks his own female, or perhaps is deceived by another male mimic. Can you suggest why all scorpion fly males aren't transvestites?

Sneaky copulation

In male–male contests large males usually win. Smaller males have two options: either to wait to grow larger or be a sneaky or 'satellite' male. A study of bullfrogs (*Rana catesbeiana*) found that the most successful males, usually the largest, call from the edges of the pools to attract females. However, some males adopt the sneaky tactic: they sit and wait close to calling males and try to grab females as they hop or swim by.

Although small, sneaky males gain few matings, they do gain some and so both tactics can coexist. It may be that small males learn something about competition which they can put to use in subsequent years when they are bigger. So males in the same population can behave in different ways, though the frequency of each tactic may not be the same. Small males thus become sneaky as this is the best tactic for them to use, given their size.

6.4   Female competition and selection of mates

Since females are the scarce resource we might not expect female–female competition to be as significant as male–male competition. This is generally so. Almost all females who are able to mate do so, provided that they are in good physical condition. However, female–female conflict and aggression do occur. For example, in Mormon crickets (*Anabrus simplex*) females compete for access to males. In this case males can afford to be choosy about which female to mate with since they provide nourishment as well as sperm during mating. The males not only produce a **spermatophore** (a package that contains sperm and is transferred to the female during copulation), they also give the female a spermatophyllax (an edible portion to nourish her) and this is especially valuable when the food supply is rather patchy. Under these conditions, female–female aggression is considerable as they fight over males.

In most species, however, female–female competition and conflict is limited and their greatest concern is which male to mate with. The selection seems to be primarily for good quality resources and/or good quality genes. How do females make such an assessment?

Selection for good quality resources

There are many resources which females might seek from males, two of the key ones are **territory** and **food**. In some cases, such as song birds (e.g. titmice and thrushes), the two may be inextricably linked because all food during the breeding season is collected within the territory.

Why should females prefer to mate with males that hold good quality territories? If they do, the females benefit since their offspring will usually have an improved chance of survival through greater food availability, more rapid growth and lower predation risk. Territories differ in the protection they give to the territory holders. Some territories offer good hiding places, some do not. For species nesting in colonies, like the lesser black-backed gulls mentioned earlier in this chapter, a central location in the colony is advantageous since predators approaching the colony tend to pick off the eggs, chicks or adults at the edge. Also, at the edge there are fewer birds to mob predators as they approach and so attacks by predators are often more successful.

Why should females prefer to mate with males that are good providers of food? Food is an important resource for the female, as well as for the offspring. A study of common terns (*Sterna hirundo*) found that females assess the quality of a male as a potential partner by the amount of fish he presents during the courtship period and prior to mating. The food, of course, is obtained at sea, not in the nesting area. This feeding not only builds up the body reserves of the female in preparation for egg laying and incubation but also serves as an indicator as to how good a provider of food the male will be for their chicks. Mate choice is crucial to chick survival. If a female makes an unwise choice her genetic investment may be wasted.

Selection for good quality genes

We would expect females to select males who offer good genes, that is genes which increase the survival and subsequent reproduction of her offspring. But how can a female tell that a male has good genes just by what he looks like and what he does? The actual mechanisms used by females in mate choice are not fully understood but their effects can be demonstrated.

A study of fruit flies (*Drosophila melanogaster*) set out to determine if a female can select a male with good genes, in this case, genes that produce competitive offspring. In this laboratory study a batch of females flies were divided into two groups. One group was given the opportunity to select a mate, the others were not allowed to choose but were randomly assigned a mate. After the eggs laid by the females had all hatched, the researcher allowed the larvae that emerged from the eggs to compete with each other for access to food. It was found that the competitive ability of the larvae of the females who had selected a mate was better than the larvae of females who were not given a choice. It follows that females, in ways we do not yet appreciate, are able to select a male whose genetic qualities give a competitive advantage to their offspring.

Mate choice by females is believed to explain why the males of some animal species possess features that are strikingly different from the female. Charles Darwin was the first biologist to put forward a theory as to why male and female animals of the same species, peafowl for example, often vary in appearance. His idea was the **theory of sexual selection**. Darwin believed that sexual selection always leads to the same two outcomes, namely, males compete with one another to be selected by females and females select a male to mate with. The outcome of female choice sometimes produces males with very exaggerated features, ones that might appear to handicap the animal.

Evidence for Darwin's theory has been provided by a study of an

African bird. The long-tailed widow bird (*Euplectes progne*) is a fairly common bird that inhabits grassland in Kenya. For most of the year, the female and male birds are black and females have a tail about 7–8 cm long. However, in the breeding season the males acquire red breast feathers and a very long tail. The birds are some 15 cm in length (from beak to the base of their tail) but their tail is about 50 cm long. The long tail is easily seen by females as the males fly around their territory trying to attract females. Males may mate with several females in the breeding season.

In one study of long-tailed widow birds in the field a researcher caught a number of male birds and divided them into four groups:

- group A had their tails cut in half;
- group B were caught and later released without change to their tails;
- group C had their tails cut off and then glued back, restoring them to their original length;
- group D had their tails lengthened by gluing on the extra half tail from the birds in group A.

The males from all four groups were then released and the number of nests that were built in the territory of each male was counted. The result was very clear: males with lengthened tails attracted more females, those with shortened tails had the least success (figure 6.6). Females thus select males with long tails, tail length being an indicator of male quality. Long tails have an adaptive function, in spite of being an apparent handicap.

More recently, other theories have been proposed to explain sexual selection, and the study of the widow birds is a useful one to use to consider how two competing theories, the runaway theory and the handicap theory, explain tail length in males.

Runaway theory

The central idea of this theory is that female choice can be made on the basis of a character that is arbitrary. It might be a long tail, as in widow birds, but it could be some other character. If these physical characters are inherited, as well as being selected by females, and especially if they give their owner some immediate advantage in male–male competition, then these characteristics are passed on to the female's sons. Their sons will, in turn, be selected as mates by females. This process, when repeated, allows these genes to be passed on. As a consequence, sexual selection tends to favour males with longer tails, provided there are reproductive benefits to the possessors.

This is called the runaway theory since selection continues to lengthen tails until the cost of possessing them, such as difficulty in flying, is greater than the potential benefit of selection by females.

Handicap theory

This theory offers a different explanation. It suggests that females select a male with a long tail simply because he possesses an attribute that could be potentially disadvantageous. So if a male does have a very long tail this is an

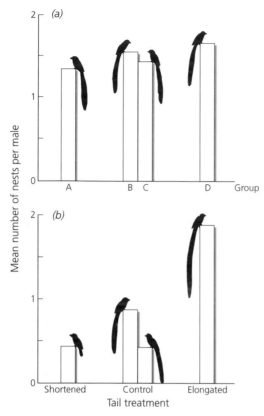

Figure 6.6 Mating success in male long-tailed widowbirds subjected to different tail treatments. (*a*) Mean number of active nests per territory for the nine males of the four treatment categories, before the experiment. (*b*) Number of new active nests in each territory after treatment of males.

indicator to a female that he is a good survivor. If this is so, and this trait can be passed on via his genes, then their sons will be good survivors: females choose such males because they can demonstrably overcome this handicap.

6.5   Female–male conflict

The cause of conflict between the sexes is due to the difference in the reproductive effort of each sex: males generally put their reproductive effort into mating, females into parenting. The resulting conflict shows itself in a number of ways.

Rape
Some people think that rape only occurs in humans. To an ethologist, though, rape, or something very like it, seems to happen in a number of

species. For example, in mallards (*Anas platyrhynchos*) the response of a male paired with a female to the appearance of another male is aggression as he attempts to drive off the intruding male. However, if the paired male is temporarily absent and then returns to find a male copulating with the female, his response may be to rape her. By doing so he ensures that his sperm compete with the first male's sperm.

Infanticide

Infanticide is the killing of offspring and it occurs in mammals, usually when a new male, or males, takes over a group of breeding females. There is obvious conflict here since it is not in the interests of the females to have their offspring killed. However, it is in the interests of the new male(s) to kill the existing young since this brings the females into breeding condition, so that they are ready to mate and produce young that have the genes of the new male(s) sooner than would otherwise be the case. This behaviour has been observed in lions (*Panthera leo*) and langur monkeys (*Presbytis entellus*).

Multiple mating

Multiple mating occurs when either a male mates with several females or a female mates with several males. A conflict arises here since females usually have little to gain by multiple mating: for them it is better to mate with one, the best available. However, for males multiple mating is clearly advantageous since by doing so they increase their chances of producing offspring.

Although, in general, it is not in the interests of females to opt for multiple mating there could be some circumstances in which it is beneficial. Researchers have suggested a number of possible explanations for this:

- if each male provides her with food during copulation she can build up her body reserves and produce more offspring;
- several males could help to rear the offspring, which could be important in social animals;
- it could serve as insurance, in case one male is sterile;
- it generates genetic variation among offspring, which may be advantageous;
- in some species one male may not provide enough sperm to fertilise all her eggs.

Parental care

For a female animal, fertility is limited by the number of eggs she produces; for a male, fertility is limited by the number of inseminations he obtains over his reproductive lifetime. Since females produce far fewer eggs than males produce sperm, the competition amongst sperm to fertilise eggs is severe. In most species, therefore, each male attempts to mate with as many females as possible.

If only one parent is needed, and usually this is the female, then it generally pays the male to leave after mating and seek other sexually receptive females. Females are literally left holding the babies while males desert. However, if it requires two parents to rear the young, as in most species of bird, then both males and females will, in general, need to be equally attentive to the needs of the offspring. They often tend to be similar in appearance too, as in lesser black-backed gulls which we mentioned earlier in the chapter.

6.6   Mating systems

If the courtship signals are appropriate and both partners are in a sexually receptive state, then mating occurs. There are four main kinds of mating systems: **monogamy**, **polygyny**, **polyandry** and **promiscuity**.

Monogamy

Monogamy is the system in which each breeding adult mates with only one individual of the opposite sex. This system is the norm in many bird species as it frequently requires both parents to provide food and care for the offspring after the eggs have been incubated and hatched.

When monogamy is the system that is practised, it is usually **seasonal** – that is the pair bond exists for a specific breeding season, after which the two individuals separate. The pair bond is important because it ensures that the pair act in concert, as a team: this is clearly evident in lesser black-backed gulls (section 6.1).

Monogamy can also be **perennial** – that is the partners stay together beyond the breeding season, in some cases for the duration of the breeding life of the individuals. This is particularly the case for larger, longer-lived birds such as swans. This not only facilitates the bonding between the two individuals but also shortens the courtship period after the first breeding season.

In the wild it is possible that one of the parents could be killed, injured or taken by a predator. Could the other parent cope on its own? In one study, the effects of male removal on the parental care of female white-throated sparrows (*Zonotrichia albicollis*) was investigated. The researchers removed the experimental males from their territories when the chicks were six days old. The females responded by increasing the rate at which they fed the young. Although the females did rear the young, the chicks were lighter in weight compared to young birds from a set of control nests. This shows that the female sparrows can compensate to some extent for the loss of a male, though two parents do give their young an advantage as chicks that are heavier at fledging have a higher survival rate.

In monogamous systems the male often remains close to the female. If he stays close to her he can be fairly certain that the young he helps to raise are his. If he leaves the female alone, or if she goes onto an adjacent territory, it is possible that she will mate with another male: if so, the territory holder will be helping to rear and protect young that are not his. It is now fairly easy to check the paternity of offspring using DNA fingerprinting. The technique is based on the degree of variation in the genetic material. Since the variation is so great, each individual has a unique DNA fingerprint and paternity can be determined quite easily.

A potential problem with animals practising monogamy is **desertion**. For each partner, the crucial question to ask is 'Will I leave more offspring if I leave now and find another mate?' If the answer is yes, then it pays that individual to desert. The reaction of the other partner is also important; if they also desert, then their offspring may die and the inclusive fitness of each partner will probably fall. The timing of desertion is also important. If it is late in the breeding cycle, then the partner that is left may be able to rear the brood successfully: if it is early in the breeding cycle the partner may not be able to compensate for the loss of their mate.

A fascinating situation occurs in the snail kite (*Rostrhamus sociabilis*). The birds are monogamous but either sex may desert. Desertion is more likely if food is abundant. Can you suggest why this is?

Polygyny

In polygynous systems one male mates with several females. Obviously the male thus increases his genetic contribution to subsequent generations since he is the father to a large number of offspring. Of course, this is only an advantage to successful males, that is ones that mate: some males won't mate at all! However, there are costs to polygyny, even to a successful male. One important cost is that the male is wholly dependent on the parenting skills of the female if the offspring from their mating are to survive and then reproduce at a later stage.

A feature of polygynous systems is that there is often strong **sexual dimorphism**, that is the two sexes differ in appearance, as is the case in peafowl, red deer and long-tailed widow birds. This is because there is strong competition between males for mates and because females have to choose between males on some obvious quality.

Polygyny is common in mammals, possibly because only the female is able to provide milk for the offspring. In many instances the male leaves the female to rear the young by herself, as in the tiger. Polygyny also occurs in some birds, for example, great snipe and many game birds, such as pheasant. In such species the chicks leave the nest soon after hatching and they feed themselves. The female offers them protection and leads them to food. Polygyny seems to result in female choice where males differ greatly in quality.

In some species both monogamy and polygyny can occur, for example in male indigo buntings (*Passerina cyanea*). These are sparrow-like North American birds with vivid, deep blue feathers. The males set up territories which vary considerably in quality. One study found that 10% of males had two mates, most had one mate while some went unmated. So it could pay a female to mate with an already mated male in order to enjoy the advantages of a good quality territory, for example one with a rich food supply. The study found that females who mated polygynously reared as many chicks as females who mated monogamously. So it seems that it is better for females to choose to pair with a male who already has a mate, rather than a male who doesn't have one! This is because although the male with a mate may give the second female less attention, she can still get enough food to feed the chicks because the territory is so good. Males with two mates had the best quality territories, those with no mate had the worst.

Polyandry

In a polyandrous system a female will mate with a number of males. As a consequence it is generally the male who is responsible for rearing the young, though the female may help to rear the young from the last mating in the breeding season. Polyandry is rather rare, though it is found in some bird species, especially those nesting in the Arctic tundra. In this habitat there is an abundance of summer food and feeding is possible 24 hours a day in summer, which allows very rapid egg production in females. One study of the spotted sandpiper (*Tringa macularia*) found that females may lay 3–5 clutches of eggs in a breeding season. By doing so the female significantly adds to her genetic contribution to future generations.

Promiscuity

In this system males and females mate with many different individuals. When this is practised, parental care may be minimal or may be provided by either the male or the female. There is no pair bonding between the two individuals, but selection of a mate is still important, especially for the female. In spite of apparent advantages of this system to males – for example the male does not have to bring up the offspring – promiscuity is not common.

6.7   Human courtship and reproduction

Humans are subject to the same evolutionary forces that operate on other animals, namely the need to survive and reproduce. The desire to mate and reproduce is very strong in most humans, and courtship behaviour plays an important part in our lives, especially in adolescence and early adulthood.

An important difference between human animals and non-human animals is that humans may be able to control the outcome of a mating event through the use of contraception. When other animals mate, the unconscious 'desire' of both female and male is to produce offspring. (This does not mean that copulation in animals may not serve some other function too – in bonobo chimpanzees it is thought to strengthen social bonds between the individuals in the group.) In humans frequent copulation – which is possible since the human female is sexually receptive throughout the year – also strengthens the ties between a couple. This may be crucial for the functioning of the couple as a unit, especially if they have children. Intercourse is a very powerful social cement.

The fact that human males are taller and heavier than females suggests that competition between males occurred in our evolutionary past, though this may be less critical today. This sexual dimorphism was the result of strong selection through female mate choice. Nowadays, human strength and weaponry may be seen in financial rather than physical terms. Males who have large bank balances may be particularly desirable as mates, whether or not they are tall and muscular.

Intersexual selection occurs in humans. Males may give considerable attention to their physical appearance in order to be attractive to females. This can be readily seen in the amount of time and financial resources allocated to grooming and clothing. In most societies, though, females spend more time than males on such activities. This suggests, as we all know, that in humans, unlike most other animals, males are as choosy about their mate as are females (figure 6.7).

Figure 6.7 Mixed-sex activities, such as dancing, serve a number of functions in human society. One of them is to help both females and males identify the strengths and weaknesses of possible partners.

How do humans choose a mate?

For most humans there is quite a large number of potential mates. Social taboos rule out some possible partners, particularly close relatives such as sisters, brothers, fathers and mothers. There is a good biological reason for this, namely the danger of **inbreeding** (that is breeding with a close relative). Inbreeding increases homozygosity and therefore the chance of offspring having congenital deformities or being weak and sickly and thus more likely to die. Indeed, in all existing human societies it is illegal to marry such relatives. This is an example of a social convention existing for good biological reasons.

In choosing a potential partner most people take a number of factors into account. Three that have been investigated are physical attractiveness, similarity and proximity. Studies of the 'Lonely Hearts' sections of newspapers and magazines have provided evidence of the importance of these, and other, factors in human courtship. (See figure 6.8 for an indication of these three factors in influencing attraction for females and males.)

Physical attractiveness is of overwhelming importance and especially in initial attractiveness. The large amounts of money and time people spend on diets, cosmetics, plastic surgery, and so on are eloquent witness to this.

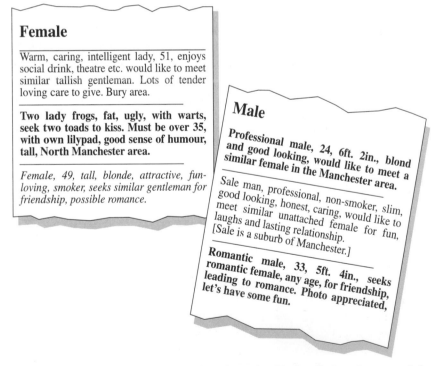

Female

Warm, caring, intelligent lady, 51, enjoys social drink, theatre etc. would like to meet similar tallish gentleman. Lots of tender loving care to give. Bury area.

Two lady frogs, fat, ugly, with warts, seek two toads to kiss. Must be over 35, with own lilypad, good sense of humour, tall, North Manchester area.

Female, 49, tall, blonde, attractive, fun-loving, smoker, seeks similar gentleman for friendship, possible romance.

Male

Professional male, 24, 6ft. 2in., blond and good looking, would like to meet a similar female in the Manchester area.

Sale man, professional, non-smoker, slim, good looking, honest, caring, would like to meet similar unattached female for fun, laughs and lasting relationship.
[Sale is a suburb of Manchester.]

Romantic male, 33, 5ft. 4in., seeks romantic female, any age, for friendship, leading to romance. Photo appreciated, let's have some fun.

Figure 6.8 Examples of adverts taken from the 'Perfect Partners' pages of the Manchester Metro News, 24 March 1995. The adverts illustrate the importance of physical attractiveness, similarity and proximity in selecting potential partners.

Similarity suggests that people generally look for a partner of the same ethnic group, social class, intelligence, religious convictions and personality. By and large, most people are attracted to, and eventually marry, others with similar backgrounds. Dating agencies also use this principle when matching couples. **Proximity** is also important because we are more likely to find a sexual partner, or simply a friend, from those who are near to us. Students in their first year at university are more likely to make friends with those in the rooms near to them.

Courtship between humans offers the opportunity for potential mates to demonstrate their worth. In human terms this may mean fidelity (that is sexual faithfulness), the ability to obtain valuable resources (such as a job) and generosity. A considered choice of partner is especially crucial for females. This is because, if the couple split up, any children usually remain with the mother.

Male–female conflict is seen in humans, as in other animals. This is partly because, as a generalisation, male effort is largely put into mating, the female's into parenting. Rape is almost always carried out by a man on a woman or girl (though some men rape other men or boys). However, rape only very rarely (one or two cases in a thousand) leads to pregnancy. Some biologists believe that rape is a pathological behaviour and has never had an adaptive function. Other biologists believe that rape may once have served as a way of males slightly increasing their reproductive success.

Infanticide has been practised by some societies for a very long time. Infanticide does not require that one, or both, of the parents actually kills the child; it is enough for a child to receive little food or protection or to be abandoned. Usually daughters are more likely to die in this way than are sons. This is because males are often perceived as having a higher social value. For example, sons may inherit land, while daughters may be expensive to get married. The effect is to limit population growth. This is the hope of the Chinese government. They have a single child per family policy coupled with female infanticide, which limits, or at least slows down, the rate of increase in the population.

Conflict over parental care is seen in humans and it is usually the female who ends up caring for the offspring. In the United Kingdom, a significant percentage of families are single-parent families (in 1998 this was approximately 20%). This reflects both a change in attitude towards lone parents and the fact that economic conditions and social support systems mean that one person can now rear a family on their own, albeit often with considerable difficulty.

Male–male conflict is evident in humans and mate guarding of females by males is a reflection of this. It is possible that mate guarding by males of females is more important nowadays as females are more likely to be out working and thus spend less time in the home. At work females come into contact with potential sexual partners and so mate-guarding tactics are probably encouraged by males. In reality though, there is little a male can do to prevent a female, when out of his sight, choosing to mate with another male. Genetic tests show that in the United Kingdom some 10% of children born in wedlock are not the children of their 'father' (that is their mother's husband). Traditionally, one way in which people signal their unavailability to potential mates is by wearing a wedding ring. Interestingly, the trend in recent years has been for wedding rings to be worn not only by females – as was once the case – but also by males.

Female–female conflict might also be expected in humans. Courtship is a time of assessment and it is in the interests of the female to prolong it without having unprotected intercourse until she is sure that the male is 'ideal' for her. For females, testing the genetic qualities of males is not easy. Perhaps the absence of obvious deformities or susceptibility to disease is an indication of adequate genetic quality. Human females do compete for mating opportunities with good quality males. Such conflict may simply manifest itself in that good quality males (whether of high genetic quality or with significant environmental resources) are more likely to be snapped up young than are poor quality males.

The main human mating system is monogamy, with **serial monogamy** – in which a person has a number of such relationships during their life, each one of which is monogamous – becoming increasingly common. There are some societies that practise polygyny, for example the Kipsigis tribe of south-western Kenya. However very few practise polyandry, though this is found in some Tibetan groups. Of course, this case is quite unlike that of the sandpiper example mentioned earlier. In Tibet there is no food glut in summer, or at any other time of the year. The farmers, and the farming regime, there are so poor that it would be difficult to support a family if each had a wife: so each man has to 'share' a wife.

Monogamy may not have been the norm when we were hunter-gatherers since good hunters could ensure a better food supply and could support more than one mate. However, once societies could produce a food surplus and lived in villages, towns and cities monogamy seems to have become more common. The move towards monogamy was probably aided by the fact that a human child requires parental care for a very long time before it completes its development at the end of adolescence. Though it is clear that both males and females may be sexually unfaithful, most parents remain together. This indicates that both parents are prepared to pay the high price of raising their children until they are mature, independent and capable of reproducing, thereby increasing the inclusive fitness of their parents. The inclusive fitness of the parents takes into account not only their own children but also the offspring of their children and other kin too.

So far we have said nothing about homosexuality or celibacy. **Homosexuality** is when a person feels sexually attracted mainly towards people of the same sex. Many homosexuals, whether women or men, end up having children. Others, though, never have children of their own. It is difficult, therefore, to see how homosexuality could be a product of natural selection, though it has been suggested either that homosexuals help bring up their relatives (such as nieces and nephews) or that homosexuality is linked with some other advantageous characteristic such as creativity.

In much the same way we are unsure whether **celibacy** (in which a person does not engage in sexual intercourse) is a product of natural selection. Clearly, if an only child remains celibate then it reduces that person's inclusive fitness. However, if the celibate has two extra siblings, four extra nieces or nephews or eight extra cousins, their genetic endowment, in theory at least, is met by the reproductive activity of these extra relatives.

Of course, just because homosexuality and celibacy may not be advantageous in evolutionary terms tells us nothing about whether they are right or wrong. In some cultures homosexuality and celibacy are thought unacceptable. In others they may be associated with a particular role in that society, or it may be felt that such issues are a matter for the individual concerned.

SEVEN

Social behaviour

Species vary in the degree to which the individuals in them are social. At one extreme are the social insects where a colony consists of a large number of individuals all apparently co-operating for a common good. At the other extreme are species where individuals live almost all of their lives in isolation. However, even these species show a certain amount of sociality in producing offspring. In this chapter we begin by looking at parental behaviour and then go on to consider richer instances of social behaviour.

7.1   Parental behaviour

In chapter 6 we looked at how natural selection has shaped the courtship and mating behaviour of animals. However, natural selection does not end once eggs have been fertilised. Species vary greatly in the amount of parental investment that individuals give to their offspring. As we saw in section 6.5, males and females often differ in the amount of their **parental investment**. In the majority of mammals females provide most of it, though in some monogamous species, for example gibbons, males also play a significant role. As we saw in section 1.4, sticklebacks are an example of a species where the female provides no parental investment once the eggs have been laid. Instead the father incubates the young and aerates the nest. He even protects them from predators for a while after they have hatched.

So why are there some species where females provide most of the parental care (such as most mammals), some where males provide most of it (for example sticklebacks) and some where both sexes provide roughly equal amounts (such as gibbons and many birds)? From an evolutionary perspective it is best to look at the options a parent has. At its simplest, these are either to desert or to continue to invest in the offspring. In species with internal fertilisation the male has the opportunity to desert before the female. In species with external fertilisation, the opposite is the case. (Think about it!)

Perhaps surprisingly, the simple prediction that arises from this fact fits much of the data. Mammals and birds have internal fertilisation and the male almost never looks after the offspring on his own whereas the female often does. Some fish have internal fertilisation and some external. Female care is commonest with internal fertilisation and male care with external. However, there are many exceptions and there is more to being a parent than simply not deserting. It turns out that in fish, care by the father is correlated more closely with male territoriality than with external fertilisation. Perhaps defence of a territory for the purposes of mate attraction can easily evolve into male care of the young as the territory serves as a safe home for the young. This would mean that male territoriality provides a **pre-adaptation** for male care.

In some mammals, care of the offspring by the mother lasts for many years – chimpanzees, elephants and humans are notable examples. Such extended care provides greater opportunities for offspring to learn from their mothers. Offspring can learn how to forage more effectively, how to avoid predators and how to interact socially with other members of the group. Poaching of African elephants for ivory has led to a number of cases where young bull (male) elephants live in the wild as orphans, their mothers and other older family members having been killed for their tusks. These orphaned bull elephants have been described by game wardens and local people as 'delinquents'. They are more likely to attack people and have even been found trying to mate with rhinoceroses, raising the possibility that their sexual imprinting may have been interrupted (see section 2.3).

7.2   Costs and benefits of being social

Sociality, that is, living in groups, has both benefits and costs for the individuals in such groups. It is the relative balance of the advantages and disadvantages of sociality for these individuals that determines the extent to which a particular species shows social behaviour.

The main *advantages* to being social are as follows.

- For prey species that avoid predation by escape rather than by camouflage or antipredatory aggression, early detection of a predator may mean the difference between life and death. For example, as we discussed in section 5.6, the larger a flock of wood pigeons, the less likely a goshawk is to succeed in killing one of them when it attacks. The reason for this is that the more pigeons there are, the more likely it is that one of them will spot the attacking goshawk while it is still far enough away for the pigeons to take off and make their escape.
- Groups of prey may be able to fight off or deter a predator. The social insects, such as ants, bees and social wasps, can defend their colonies against predators many times larger than they are. The tussles that can develop between predators and groups of prey can be epic. A wolf pack may spend seven days tracking a herd of bison, waiting for one of the bison to separate from the pack so that they can surround and attack it.

- If a predator has been neither detected nor deterred, it still has to select a prey animal and attack it. It has been shown experimentally that in some prey species, for example certain species of fish, predators such as squid and pike do less well when attacking groups than when they attack individuals. This is because the predators become confused by potential prey darting in all directions (section 5.6). The predator shifts from one potential prey individual to another, giving all the prey a better chance of escaping. This is known as the **confusion effect.**

- Even when predators attack large groups as successfully as small groups or individuals, it may still pay the prey to live in groups (section 5.6). Imagine a group of ten prey individuals and another of one hundred. If the two groups are equally likely to be successfully attacked, and the successful predator takes just one individual, then from the point of view of each individual in the group with 100 prey, there is ten times the chance of surviving the attack than in the group with ten prey. This effect is sometimes referred to as the **dilution effect**. The dilution effect is extremely important in nature. It not only explains why many prey live in groups but also provides at least one reason why it may pay an individual to be born at the same time as other individuals, in a concentrated birth season. An example of this is shown in figure 7.1. The more mayflies that emerge on an evening, the less likely each mayfly is to be taken by a predator.

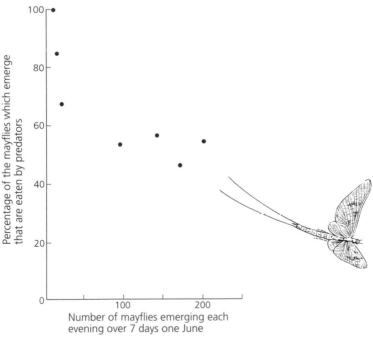

Figure 7.1 In mayflies, it pays to share your birthday. The more mayflies that emerge on a particular evening, the more likely each of them is to escape the clutches of predators.

- Predators too benefit from living in groups. Many social carnivores are able to catch larger prey when they co-operate than when they hunt on their own. This is the case for wolves, lions, spotted hyenas and African wild dogs (also known as hunting dogs).
- Being in a group enables you to learn from others in the group. This may simply involve following group members to obtain food, water or shelter, which is known to occur in some bird flocks. Or it may mean learning much more than this. Many social animals have a **culture** which they learn from other group members as they grow up. Depending on the species, this may include how to form alliances, how to fight or how to use tools.
- If there are enough animals in a group, they may make the physical environment more tolerable. Woodlice can reduce the risk of desiccation by aggregating. Penguins may huddle for warmth. Colonies of social insects can maintain themselves at temperatures close to 30 °C, meaning that, to all intents and purposes, they are as homeothermic as some mammals and birds. Certain shoals of fish swim in ways that reduce drag while some birds, such as geese, reduce drag by flying in formation.
- Finally, there are advantages which follow provided the group members belong to different species. For example, groups of impala and olive baboons reduce the risk of predation by combining the sensitive noses and ears of the antelopes with the colour vision of the monkeys, which may sit high in trees or cliffs on the look out.

However, there are *disadvantages* to living in groups.

- The more animals there are in a group, the more food the group needs. This may not matter too much in grazing animals such as rabbits, deer or antelope, but it undoubtedly puts a severe restriction on the number of individuals there can be in a pack of social carnivores such as lions, hunting dogs or wolves since the pack members compete for food.
- There may be competition within the group for resources in addition to food. For example there can be competition for mates and subordinates may find themselves unable to reproduce. This is particularly true for males. In many polygynous species, mating is dominated by only a small proportion of the males. In some species, dominant females may also monopolise reproduction (see section 7.5).
- Diseases generally spread more easily when there are more of you. This is because transmission from an infected individual to an unaffected one is generally more likely if there are many animals in a group.

The fact that we can list more advantages than disadvantages for group living should not lead you to think that group living is always advantageous! In some species the disadvantages outweigh the benefits – such species are solitary. Even in social species, there comes a point with increasing group size where the advantages and disadvantages balance out. It would not be advantageous for individuals to be in a group larger than this.

7.3   Evolution of altruism

In section 1.5 we introduced the idea that there are times when an animal's behaviour decreases its individual fitness but benefits another individual. This is known as **altruism**. The question is, how does altruism evolve? One way of beginning to answer this question is to ask the apparently rather trivial question 'Why do parents invest in their offspring?'. The answer, of course, is that all individuals eventually die, whether from senescence (old age) or some other cause. Individuals therefore have to reproduce and invest in their offspring to try to, as it were, keep copies of themselves alive in future generations. The optimum amount of parental investment varies from species to species and may differ between the father and the mother. After all, the ideal is when you can get another individual to look after your offspring! This is what happens in species where only one sex looks after the offspring. As we have seen, it is more likely to be the mother that looks after the young though there are species where the opposite is the case.

The previous paragraph looks at things from the point of view of individuals and after all, it is individuals we see born, reproduce and die. Yet, individuals, from the perspective of evolutionary time, are ephemeral. They live out their transitory lives and then die. The genes that they carry within them last far, far longer. Something of a revolution in the science of behaviour happened in the 1960s and 1970s when increasing numbers of scientists who worked on behaviour began to think of genes in the way that population geneticists had since the 1930s. The initial impetus for this type of thinking came when people began to ask precisely when it pays an individual to help its relatives, a behaviour that can often be explained by **kin selection.**

Kin selection

Some of the most dramatic examples of kin selection occur in the social insects. In the African termite (*Globitermes sulfureus*), for example, members of the soldier caste are literally walking suicide bombs. Huge paired glands extend from their heads back through most of their bodies. When they attack ants and other enemies, they fire a yellow glandular secretion through their mouths. This congeals in the air and often fatally entangles the aggressive soldier termites themselves as well as their hapless opponents. The spray appears to be powered by muscular contractions in the abdominal wall. These are sometimes so violent that the gland explodes together with the abdomen, spraying the fluid and the soldier in all directions.

This behaviour is an extreme form of altruism towards kin. The soldier termites protect the rest of the colony and so protect not their individual reproductive interests but the reproductive interests of the king and queen termite, to whom they are related.

A full understanding of kin selection came only with the publication in 1964 of a pair of papers by W.D. Hamilton. These papers are widely regarded as the foundation stones of sociobiology. Hamilton was able to show that for altruism towards relatives to be worth it, from an evolutionary point of view, the following equation must hold:

$$rb - c > 0$$

Here, r is the degree of relatedness between the two individuals, b is the benefit (in terms of Darwinian individual fitness) that the recipient of the altruism gets and c is the cost (again in terms of Darwinian individual fitness) that the altruist incurs.

Let us consider a simple example to illustrate this equation. Suppose an animal has to choose between eating a piece of food itself and allowing its full sib (that is sister or brother) to have it. The degree of relatedness between full sibs is 0.5. According to Hamilton's equation, altruism should occur if the recipient of the food would benefit twice as much (in terms of extra offspring it would eventually produce) from eating it as the animal giving the food would if it ate it. How could this be? Well, one possibility would be if the recipient, for some reason, had not fed for quite a while. Another would be if the recipient was very young – for example, a small piece of meat might mean much more to a newly weaned carnivore than to an adult. Notice how one of the useful things about Hamilton's equation is that we are already beginning to make testable predictions.

Unfortunately it turns out to be difficult to obtain accurate values of b and c, so making precise tests of Hamilton's equation rare. It is rather easier to obtain measures of r:

- between a parent and its offspring, $r = 0.5$
- between full sibs, $r = 0.5$
- between half sibs, $r = 0.25$
- between an aunt or uncle and a nephew or niece, $r = 0.25$
- between a grandparent and grandchild, $r = 0.25$
- between two first cousins, $r = 0.125$ (an eighth).

These figures assume there is no inbreeding. When inbreeding occurs, values of r are greater than indicated above, making the evolution of altruism through kin selection easier.

Kin selection also helps provide a different angle on the phenomenon of **parent–offspring conflict**. Parent–offspring conflict occurs when there is some disagreement between a mother or father and one or more of its young. In mammals, weaning provides a good example. Offspring often attempt to continue to obtain milk from their mother for a considerable period of time after the mother has begun to try to wean them. Nothing very surprising in that, you might think. True, but kin selection provides a different sort of evolutionary explanation for such parent–offspring conflict. Namely, that in one sense the conflict is not actually between the parent and the offspring but between one set of offspring and the next. Weaning occurs when a mother wants to stop investing in one set of offspring and begin a recovery phase before investing in another set. However, the first set of offspring are related to themselves with a degree of relatedness of 1 (by definition) but to their mother's next set of offspring with a degree of relatedness of only 0.25 or 0.5 (depending on whether they are half or full sibs). But the mother is equally related to both sets of offspring with a degree of relatedness equal to 0.5. This means that mothers are expected to be equally interested in all their offspring whereas offspring are more interested in themselves than in their sibs. Again, nothing very surprising in that. But it means that when a young mammal is sucking milk from its mother, in one sense it is taking the milk from its as-yet unborn sib(s) and that is where the main evolutionary conflict lies. To a certain extent, the mother is simply piggy-in-the-middle.

An example of a species which illustrates altruism that has evolved by kin selection is the lion (*Panthera leo*). Lions live in stable social groups called prides. A pride typically consists of between 4 and 10 adult females, their offspring and a coalition of between 2 and 5 adult males. Daughters born into the pride remain there for life while sons leave before they reach reproductive maturity at the age of about three years. As a consequence, the females within a pride are quite closely related. This is probably at least a partial explanation for two features of their lives:

- they co-operate in hunting together;
- unusually for mammals, **communal suckling** occurs so that cubs in a pride suck from any adult female with milk, not just from their own mother. Communal suckling is helped by the fact that all the adult females in a pride tend to come into oestrus at the same time, so that all the cubs are born at about the same time.

Recent research has shown that although females co-operate, when a pride is threatened by females in another pride which is trying to expand its own territory, it is always the same females that defend it. Why some lionesses hang back while others do battle is still unclear. Perhaps some females are better fighters than others. There doesn't seem to be any relationship between whether a mother hangs back and whether her daughters do, so there is no evidence for a genetic basis to this behaviour.

The adult males in a pride typically only remain with the pride for two or three years. After this they are driven away by other male coalitions. When a pride is taken over by a new coalition, the new males sometimes seek out and kill as many as possible of the cubs. The function of this **infant-icide** seems to be to bring the females back into oestrus since, as is the case in many mammals, lactating females do not ovulate. Infanticide therefore results in the new males siring their own cubs more quickly than would otherwise be the case. As many as 25% of all cubs may die from infanticide. (As mentioned in section 6.5, infanticide is known in quite a number of animals including gorillas, langur monkeys, mice, some birds and some spiders. There is no one evolutionary explanation for its occurrence but it illustrates the ways in which individuals can have very different interests.)

DNA fingerprinting has shown that when there are four or more males in a coalition, they are generally all related to each other. In such coalitions, the vast majority of the cubs produced are sired by just two of the males. Kin selection is presumably the explanation why the subordinates put up with this state of affairs. However, when just a pair of males form a coalition, it transpires that the males are often unrelated. As one might predict, both males end up siring cubs. It wouldn't make much evolutionary sense for a male lion to forgo reproduction if he was unrelated to the other male in his coalition. This sort of co-operation, between unrelated individuals, has been called **mutualism**. Mutualism is when individuals co-operate because each gains increased individual reproductive success as a consequence. Kin selection is not needed to explain mutualism.

Reciprocal altruism

A further way, besides kin selection and mutualism, in which altruism can evolve is by reciprocal altruism. **Reciprocal altruism** can be summed up as 'You scratch my back, I'll scratch yours'. It differs from mutualism in that there is a delay between one individual helping another and then getting help back in return.

An elegant example of reciprocal altruism is provided by the vampire bat (*Desmodus rotundus*). These endearing creatures, as is well known, fly out at night to attack large mammals from which they draw blood (figure 7.2). If an individual fails to obtain a meal for three nights it usually dies from starvation. A bat that has failed to obtain a meal during the night is usually fed by another bat on its return to the colony. The altruist regurgitates blood for the hungry bat. Observations show that bats which have received regurgitated blood subsequently reciprocate; in other words, on another night they will regurgitate blood for the bat that helped them. It turns out that unrelated bats are just as likely to help each other as are related ones so kin selection cannot be the whole explanation: reciprocal altruism is involved.

There aren't that many species known in which reciprocal altruism has been proved, though it also occurs in dwarf mongoose, chimpanzees and vervet monkeys. These species, with vampire bats, share three characteristics: they are long-lived, they live in stable groups and they have large brains. Being long-lived and living in stable groups gives time for relationships to build up between pairs of individuals. The possession of a large brain may be important because one difficulty in the evolution of reciprocal

Figure 7.2   A vampire bat feeding on a pig.

altruism is that **cheating** can occur. A cheat is an animal that receives help but later fails to reciprocate. For cheating to be less successful as a strategy than helping, altruists must have some way of recognising cheats and then declining to help them subsequently. Individual recognition and a good memory are therefore needed – both of these characteristics presumably being associated with large brains.

7.4   The social life of insects

There are more workers in a single colony of driver ants than there are elephants, lions and tigers in the whole world. There are more species of ants in every square kilometre of Amazonian rainforest than there are species of primates in the whole world. The social insects are seriously understudied. The best researched of them are the ants, yet only around 150 of the 12 000 living ant species have had their behaviour studied in any detail.

The most extreme form of sociality is known as **eusociality**. Eusociality is characterised by:

- co-operation in caring for the young;
- reproductive division of labour with some individuals remaining permanently sterile throughout their lives;
- overlap of at least two generations that contribute to colony labour.

Eusociality is thought to have evolved on at least 12 occasions in insects: once in termites and at least eleven times in the order of insects known as the Hymenoptera which includes ants, bees and wasps. So what is it about the Hymenoptera that makes them such ideal candidates for eusociality? We still don't know. At first it was thought that a remarkable feature of their genetics provided the answer. The Hymenoptera are almost unique in being **haplodiploid**. In haplodiploid species, females are diploid but males are haploid and so possess only one set of chromosomes. This means that daughters inherit all of their father's genes but only pass on half of their genes to their offspring, whether sons or daughters. In the absence of inbreeding, the degree of relatedness between full sisters is therefore 0.75. It was thought that this explained eusociality because the workers in a colony of ants, bees or wasps are all females.

When you think about it, though, workers in a colony of social insects aren't really helping their sister workers, they are helping their mother (the queen) to produce the reproductives that start other colonies. Although the degree of relatedness between a worker and a reproductive sister is 0.75, between a worker and a reproductive brother it is only 0.25. It isn't obvious, therefore, why haplodiploidy seems especially to be associated with eusociality, though all sorts of increasingly complicated explanations have been proposed. There is a lot we still don't understand about animal behaviour.

A very different sort of explanation for sterility in social insects is that this may be an example of **parental manipulation**. Perhaps the queen is forcing the workers to forgo reproduction, possibly by her production of pheromones.

Life in a honey bee colony

The honey bee (*Apis mellifera*) illustrates many aspects of sociality in insects. At the peak of its numbers in summer, a healthy colony of honey bees contains up to 80 000 **workers**, a few hundred males known as **drones**, a single fertilised **queen** and a number of combs containing eggs, larvae, pupae and stores of honey and pollen. The workers, drones and the queen make up the three **castes** in the colony, a caste being a collection of individuals of a particular morphological type that performs a specialised function.

A young queen-to-be begins life as a larva in one of a small number of special royal cells. These cells are larger than the worker cells and hang vertically down, whereas worker cells are horizontal. On emerging, the virgin queens fight violently among themselves. The single survivor makes up to a dozen nuptial flights during each of which she mates with a different drone from a hive other than her own. Drones are attracted to her by a pheromone she produces. She then returns to her hive and is fed on 'brood-food', a proteinaceous and highly nutritious substance secreted by the brood-food (pharyngeal) glands of workers. She grows larger and within three or four days begins to lay eggs. Throughout her life, which may last several years, she is able to use the sperm from her original nuptial flights. These sperm are stored in a spermatheca and the queen is able to lay either unfertilised or fertilised eggs, as the need arises. The bulk of her eggs are fertilised and develop into females but haploid males are produced by parthenogenesis.

The queen's task in the hive is to produce the eggs that give rise to all the other colony individuals and to regulate the colony through her production of pheromones. She prevents the workers from rearing other queens by producing 'queen substance' – actually the pheromone *trans*-9-keto-decanoic acid – from her mandibular glands in sufficient quantities for each worker to receive around 0.1 μg a day. If conditions have allowed a colony to grow significantly in numbers, the queen's production of this pheromone decreases, the workers construct a small number of the special royal cells and the old queen flies off in a swarm with a large number of workers to found a new colony. The remaining workers are now led by the surviving new queen, once she returns from her nuptial flight.

Males have no function apart from trying to mate with virgin queens from other colonies during nuptial flights.

Workers only live as adults for around six weeks, except for those that overwinter with the colony. We discussed their remarkable round and waggle dances in section 3.6 when considering honey bee foraging. Foraging is a risky occupation and is only carried out by a worker bee towards the end of her life (figure 7.3). The youngest workers either clean out brood cells from which bees have recently emerged or help incubate the brood. By the time a worker is five or six days old, her brood-food glands begin to secrete and she feeds the larvae on brood-food. Young worker and queen larvae are both fed 'royal jelly', a highly nutritious food secreted by the hypopharyngeal and mandibular glands of the workers. After three days worker bee larvae are switched from this to a diet of pollen and nectar or dilute honey. The queen larvae, though, continue to be fed on royal jelly.

By the time a worker is 10 to 12 days old, her brood-food glands have greatly diminished in size. Wax-producing glands on the underside of her abdomen have become activated and she begins comb-building and repairing, which requires the production of fresh beeswax. Only when she is about three weeks old will a worker begin to forage. A few workers also help in guarding the hive against intruders. Chemical alarm communication is well developed and involves pheromones apparently especially evolved for the purpose.

We said that eusociality evolved at least a dozen times in the insects. Before we leave eusociality, it is worth mentioning that it has also evolved in

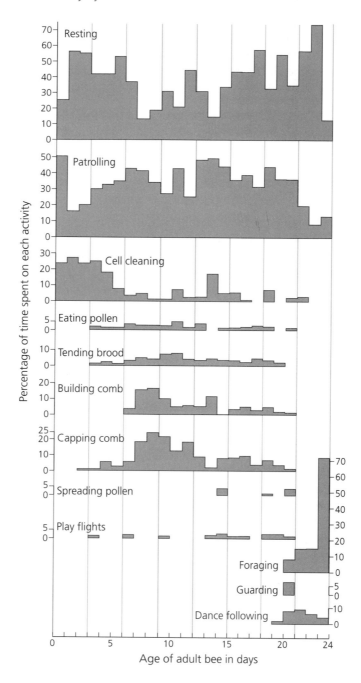

Figure 7.3 The jobs that a worker honey bee does change as she ages. This graph shows a complete record of the behaviour of a particular honey bee throughout her three-and-a-half weeks of adult life. (From M. Lindauer (1961) *Communication among social bees*. Harvard University Press.)

at least two species of mammals (in each case in the African mole rat family) and in one species of shrimp. The most studied of these three species is the naked mole rat (*Heterocephalus glaber*). These are extraordinary animals (figure 7.4). They are blind, effectively ectothermic, have practically no hair and lack a circadian (daily) biological clock. They live in colonies with typically 100 (sometimes 300) individuals. However, only one female (the 'queen') and only between one and three males reproduce. The remaining 98% of the animals in the colony never reproduce, even though naked mole rats can live for over 20 years, a remarkably long time for an animal where most adults have a mass of only 30–40 g and even the queen only weighs 60 g. Suppression of reproduction is maintained by behavioural interactions between the dominant queen and the other individuals. If the queen dies, several females come into breeding condition and fight to the death among themselves until only one is left.

So why are naked mole rats eusocial? The species lives permanently underground and feeds on subterranean roots and tubers in arid regions. By living in a large colony, naked mole rats are able to exploit an ecological niche in which a solitary animal might fail to survive because of the hardness of the soil and scattered occurrence of the tubers on which they feed. It is known that colonies are extremely inbred. We noted earlier that altruism through kin selection can evolve more easily where there is inbreeding because the degrees of relatedness between individuals are higher. Eusociality in naked mole rats is probably favoured by a combination of kin selection and manipulation on the part of the queen.

Figure 7.4   The naked mole rat. A rare example of a eusocial mammal.

7.5   The social life of African wild dogs

African wild dogs (*Lycaon pictus*) occur throughout Africa. Savannah woodland is their preferred habitat, but they are found in the southern Sahara and in the snows of Mount Kilimanjaro. Population densities as low as one pack per 2000 km^2 are not uncommon and there are probably fewer than 10 000 individuals in existence.

A pack typically has between 6 and 10 adults and up to a dozen or more pups. Hunting usually takes place around dawn or dusk, when it is cooler. The chief prey are impala, puku and Thomson's gazelle. These prey have a mass of around 20–90 kg but herbivores as large as greater kudu (mass around 300 kg) are sometimes taken. African wild dogs can run at about 60 km per hour for 5 km or more and it is this ability that enables them to run down their prey. When they close in for the kill, the attack is highly co-ordinated. Different individuals in the pack specialise in going for different regions of the body, so one animal typically attacks the nose, another the tail and so on (figure 7.5). After a successful hunt, any pups present are allowed to feed first. Pups too young to take part in the chase are fed on food regurgitated to them by any adult – not just their mother.

The adult males and adult females in a pack have separate dominance hierarchies. About once a year the dominant male and female mate. On the rare occasions when two females in a pack attempt to rear pups, the

Figure 7.5   A pack of African hunting dogs pulling down a zebra. One individual has grabbed hold of the victim's nose and another its tail while two animals are beginning to disembowel it. The dominant individuals in the pack take the most dangerous roles when hunting.

dominant female may kill the other's pups. In other words, females don't co-operate, quite the reverse. Indeed, mortality among females is higher than among males and adult males outnumber females by two to one. Unusually among mammals, it is the females that disperse from the pack which they do when aged between 14 and 30 months. Males remain in the pack throughout their lives (up to 10 years or so). This sex-specific dispersal prevents inbreeding from occurring but means that the males in the group are related to one another whereas the adult females are not.

7.6   Chimpanzee society

African wild dogs are social principally because of the benefits this has for hunting. Chimpanzee sociality has a greater range of benefits.

The chimpanzee (*Pan troglodytes*) and the bonobo (*Pan paniscus*) are our closest evolutionary relatives. We probably separated from them only some 5–8 million years ago. Until recently, the bonobo (or pygmy chimpanzee) was little studied and we will concentrate here on the chimpanzee (figure 7.6). The ethologist who has devoted her life to the study of chimpanzee behaviour is Jane Goodall. When she first arrived at the Gombe reserve in Tanzania in 1960 to study chimpanzees, the game warden who took her round made a mental note that she wouldn't last more than six weeks. She is still there, carrying out the longest ever continuous study of the behaviour of a species.

Chimpanzees are semi-terrestrial, spending 25–50% of their time on the ground. They forage during the day and build nests in trees for the night. They are omnivorous, feeding to a large extent on the fruit, leaves, bark and seeds of a wide variety of plant species. They also eat termites and ants and occasionally co-operate to kill and eat small baboons and other monkeys.

The basic social unit in chimpanzee society is a loose association of about 30–60 animals that tends to remain in the same area for many years. As in African hunting dogs (section 7.5), but unusually among mammals, it is daughters who disperse, at puberty, from the troops in which they were born. This means that within a troop, males are often closely related to one another. Although competition occurs between males, and a dominance hierarchy exists, males also co-operate: they frequently groom one another and may hunt together and share food. However, DNA fingerprinting shows that closely related males often do not co-operate any more than distantly related males. This implies that evolutionary forces other than kin selection are needed to explain patterns of male–male co-operation.

In both sexes, the young are dependent on their mothers for many years. This gives a young chimpanzee plenty of time to learn from its mother. Chimpanzee behaviour shows considerable variation between individuals, resulting at least in part from learning. This means that social interactions within a group may be complex and generalisations can oversimplify.

Figure 7.6 Chimpanzees interact in a variety of ways.

Jane Goodall was fascinated to find that chimpanzees use tools. For example they may select sticks, strip them of their leaves and use them to fish for termites or ants. At the time when this was discovered, it was thought that only humans used tools. More recently Jane Goodall and other researchers have shown that chimpanzees, again like humans, can practise deception. They may behave differently, and to their advantage, if observed by other chimpanzees. For example, individuals may pretend not to notice a small, rich food source, only exploiting it when they are confident that no other chimpanzee is watching!

7.7   Human society

We began this book (section 1.1), by suggesting that it can be a mistake to generalise too readily from the behaviour of non-humans to humans, and that it can also be a mistake to imagine that we can learn nothing of our own behaviour from the behaviour of non-humans. In section 6.7 we looked in some detail at the extent to which human courtship and reproduction could be interpreted within the same framework as that used for non-humans.

This raises a more general question. To what extent is human society shaped by evolutionary considerations? There are two errors that can be made in answering this question. One is to suppose that we are entirely free of our evolutionary heritage; the other is to assume that we are totally bound by it. Consider altruism, for example. There is no serious doubt that much of human helping behaviour can be explained by the evolution of altruism through kin selection, reciprocal altruism, mutualism and manipulation. We have seen in this chapter that each of these four mechanisms occurs in some non-humans and this is surely true of humans too. We are more altruistic towards our relatives (kin selection), towards those who subsequently repay us (reciprocal altruism), when it is simultaneously to our advantage and to that of others around us (mutualism) and when we are coerced into being helpful (manipulation). But at the same time, is this all there is to altruism in humans? Can every helping behaviour of ours be explained by these four evolutionary processes? Is there no such thing as 'real' altruism, unselected for by natural selection?

We would argue that there is such a thing as 'real' altruism. Part of what it is to be human is to be a member of the natural world, subject to the pressures of natural selection as is every other species. But part of what it is to be human is to have the capacity to transcend these pressures. It is this that sets us apart from the other 30 or so million species with which we share the planet. And one of the benefits of studying behaviour is that it can help us to discern which of our behaviours are the product of our free will and which we owe to our genetic inheritance and environmental circumstances.

Further reading and videos

Books

Brooks, F. 1992 *Animal behaviour*, Usborne.

Cardwell, M., Clark, L. & Meldrum, C. 1996 *Psychology for A level* (part 6 – Comparative psychology), Collins Educational.

Clamp, A. & Russell, J. 1998 *Comparative Psychology*, Hodder & Stoughton.

Dawkins, R. 1989 *The selfish gene, 2nd edn*, Oxford University Press.

Dockery, M. & Reiss, M. 1996 *Animal behaviour: practical work and data response exercises for sixth form students*, Association for the Study of Animal Behaviour.

Giddens, A. 1997 *Sociology, 3rd edn*, Polity Press.

Goodall, J. 1986 *The chimpanzees of Gombe: patterns of behavior*, Belknap Press of Harvard University Press.

Gosler, A. 1993 *The great tit*, Hamlyn.

Hayes, N. 1993 *Psychology: an introduction, 2nd edn*, Hodder.

Hayes, N. 1994 *Comparative psychology*, Erlbaum.

Hicks, D. & Gwynne, M. A. 1994 *Cultural anthropology*, HarperCollins.

McFarland, D. (ed.) 1981 *The Oxford companion to animal behaviour*, Oxford University Press.

Manning, A. & Dawkins, M. S. 1998 *An introduction to animal behaviour, 5th edn*, Cambridge University Press.

Martin, P. & Bateson, P. 1993 *Measuring behaviour, 2nd edn*, Cambridge University Press.

Morris, D. 1994 *The human animal: a personal view of the human species*, BBC Books.

Ridley, M. 1995 *Animal behaviour, 2nd edn*, Blackwell Scientific Publications.

Tinbergen, N. 1974 *Curious naturalists*, Penguin Education.

Journals

Animal Behaviour – journal of the Association for the Study of Animal Behaviour, 12 issues a year. ASAB Membership Office, 82A High Street, Sawston, Cambridge CB2 4HJ.

FEEDBACK – education newsletter of the Association for the Study of Animal Behaviour, 3 issues a year (free). Available from Michael Dockery, ASAB Education Officer, Department of Biological Sciences, John Dalton Building, Manchester Metropolitan University, Chester Street, Manchester M1 5GD.

Videos

Among the wild chimpanzees, 60 minutes. National Geographic Society. Available from Carlton Home Entertainment via any good video shop.

Animal behaviour: the mechanism of imprinting, 15 minutes. Uniview. Available from Uniview Worldwide Limited, PO Box 20, Hoylake, Wirral L48 7HY.

Evolution by natural selection, 50 minutes. Uniview. Available from Uniview Worldwide Limited, PO Box 20, Hoylake, Wirral L48 7HY.

Gorilla, 60 minutes. National Geographic Society. Available from Carlton Home Entertainment via any good video shop.

Stimulus response, 33 minutes. ASAB. Available from ASAB Membership Office, 82A High Street, Sawston, Cambridge CB2 4HJ.

The behaviour of the 3-spined stickleback, 15 minutes. Available from Oxford Educational Resources, PO Box 106, Kidlington, Oxford OX5 1HY.

Vigilance behaviour in barnacle geese, 20 minutes. University of Glasgow. Available from ASAB Membership Office, 82A High Street, Sawston, Cambridge CB2 4HJ.

Index

altricial young 20, 22
altruism 100–4;
  reciprocal 103, 112
anthropology 2
anthropomorphism 1–2
ants 55, 68, 104–5
appeasement 39
artificial selection 14–15
attachment 19–20;
  in humans 20

bird song 15–16, 23, 33–4
breeding *see* reproduction

camouflage 32, 69
celibacy 95
cheetah 24–5, 65–6, 67
chimpanzee 39, 43, 52, 110–11;
  language 43;
  using tools 27, 52, 53, 57–8
colony 101, 104, 108;
  gull 4–6, 74, 83–4;
  honey bee 9, 40, 105–7;
  naked mole rat 108
communication 31–2, 38–43;
  honest 32, 36, 39;
  in bees 40–2;
  in humans 43
competition 8, 64, 77, 99;
  female–female 8, 83, 94;
  female–male 86–8, 93;
  human 91,93–4;
  male–male 77–83, 85, 89, 93, 110
conditioning 46–50, 53–4, 56;
  classical 46–8, 53;
  operant 48–50, 54
conflict 38–9, 83, 86–7, 93–4;
  parent–offspring 101–2
courtship 4–5, 7–8, 10–12, 30, 75–6;
  human 93–4
culture 2, 37

Darwin (Charles) 9, 84–5

displacement activity 31
dominance hierarchy 39, 64, 109–10

eggshell removal 4–6;
  cost/benefit 5
environment *see* stimulus
ethology 2, 6
eusociality 104–8;
  in insects 105–6;
  in mammals 106, 108
evolution 2, 9, 77, 101–3;
  of courtship 10–12

feeding strategies 65–6
fitness 8–9, 100;
  inclusive 9, 94;
  individual 101
food 35–6, 40, 59–68, 74, 84;
  caches 55–6, 67–8;
  selection 67–8
foraging 45, 53, 59–60, 64–6, 68–9,
    106–7;
  influences on 64–5;
  in spiders 62–3, 66;
  optimal 60–3

geese 17–18, 20–1, 22–3;
  egg retrieval in 17–18;
  imprinting 20–1, 22–3
genes 9–10, 84, 100, 105
genetics 3, 9–10, 13–15, 17, 75–6
Goodall (Jane) 110–11
groups 37, 66, 68–9, 71–3, 77, 102–3;
  advantages 68, 97–9;
  disadvantages 99;
  size 71–3, 100
gulls 4–6, 16–17, 74, 83–4;
  bill pecking 16–17;
  eggshell removal in 4–6

habituation 45–6
haplodiploidy 105
harem 77–9

homosexuality 94–5
honey bees 9–10, 40–2, 66, 105–7;
    waggle dance 40–2
human 1–3, 20, 29–30, 31, 37, 112;
    communication 43;
    courtship 90–4;
    territory 37

imprinting 20–3;
    filial 22;
    in birds 20–2, 23;
    in shrews 22;
    sexual 22–4, 97
inbreeding 24, 92, 101, 108
infanticide 87, 93, 102–3
instinct 16, 17–18, 45
intelligence 56–8

kin selection 9, 100–3, 108, 112

language 40–2, 43
learning 17, 44–58, 65, 99;
    insight 52, 54;
    latent 51–2, 54;
    observational 53–4
lek 79–80
Lorenz (Konrad) 17–18, 20–2

macaque 1, 39, 52–3;
    on artificial mother 19–20;
    using insight 52–3
marginal value theorem 61
mate choice 11–12, 74–5, 80–1, 84–6,
    89–90, 91–3
mating *see* courtship *and*
    reproduction
mating systems 88–90
maze 13–14, 51–2
memory 54–6, 104
mimicry 69–70, 82
monogamy 4, 88–9, 94;
    serial 94
motivation 29–30
mutualism 103, 112

natural selection 8–9, 32–3, 63, 69, 74,
    112
nature/nurture 13–17
nest building 7, 14–15, 31

parental care 19, 77, 87–8, 96–7, 100
parthenogenesis 75, 106
Pavlov (Ivan) 46–8
pheromones 29, 105–6

play 24–7;
    costs of 27;
    functions of 24–7
polyandry 90, 94
polygyny 89–90, 94, 99
precocial young 20
predation 6, 64–73, 84, 99, 109;
    avoiding 68–73, 97–8
psychology 3

rape 86–7, 93
reinforcement 48, 49–51;
    negative 51;
    positive 50
reproduction 33–4, 75–7;
    asexual 75;
    cost of sexual 76;
    in birds 5, 30–1;
    in sticklebacks 7–8
reproductive success 9
ritualisation 39

satellite 36, 82–3
sexual dimorphism 89, 91
sexual selection 77, 84–5
Skinner box 49–50
social behaviour 3, 68–9, 96–112
    see also groups *and* colony
social insects 97, 100, 104–7
    see also honey bees
sociality 97
sociobiology 3, 101
sociology 3
sperm competition 81
stereotyped response 7, 18
stimulus 15, 21–2, 28–9, 45–8, 62;
    conditioned 47;
    detection 15, 28–9;
    sign 7, 18;
    unconditioned 47;
    visual 21, 62–3

territory 4–5, 7–8, 31, 33–8, 97;
    cost/benefits 38;
    food 35–7, 83;
    human 38;
    mating 33–6, 74–5, 79–80, 83, 90;
    size 36, 38
tools 27, 52, 56–8, 111
Tinbergen (Niko) 3–8, 16–17
tit 33–4, 44–5, 53–4, 55, 60

waggle dance 40–2
Wilson (E. O.) 3